INFLUENCE EMPIRE

INFLUENCE EMPIRE

INSIDE THE STORY
OF TENCENT AND CHINA'S
TECH AMBITION

■ ■ ■

LULU YILUN CHEN

HODDER &
STOUGHTON

First published in Great Britain in 2022 by Hodder & Stoughton
An Hachette UK company

1

A CIP catalogue record for this title is available from the British Library

Hardback ISBN 9781529346855
Trade Paperback ISBN 9781529346862
eBook ISBN 9781529346879

Typeset in Adobe Jenson Pro by Manipal Technologies Limited

Printed and bound in Great Britain by Clays Ltd, Elcograf S.p.A.

Hodder & Stoughton policy is to use papers that are natural, renewable and
recyclable products and made from wood grown in sustainable forests. The logging
and manufacturing processes are expected to conform to the environmental
regulations of the country of origin.

Hodder & Stoughton Ltd
Carmelite House
50 Victoria Embankment
London EC4Y 0DZ

www.hodder.co.uk

CONTENTS

INTRODUCTION

It's 2002 and I'm graduating from junior high.

The final weeks of the school year were drawing to a close – as was an important chapter of our lives. We spent the last school days before summer vacation unsupervised in classrooms, discussing the future with the carefree solemnity that masks the insecurities, fears and wordless anticipation of all teenagers embarking on early adulthood.

Much of the talk revolved around our plans for the upcoming summer, the prestigious new schools we were headed to, even extracurricular classes to prepare us for the grueling three years of senior high. And many of us instinctively understood how this transition would test the bonds we'd forged during our tumultuous adolescent years.

Like everyone, I carefully curated my aquamarine yearbook, then coyly shoved it in front of close friends, casual acquaintances, teachers, and the boy I had a crush on. They jotted down their names, hobbies, and horoscopes. Most left phone numbers. But the more trendy amongst them attached a line of digits they called their QQ number.

I had no idea what QQ was – few did back then, when Microsoft Windows was still a clunky, buggy program. PCs were the preserve of the well-to-do, at least in China. Steve Jobs had yet to conceive of the iPhone. Computers were something that we had access to for 40 minutes a week at school – already considered a privilege compared with other students in Beijing, let alone across the country. The dot-com bubble was a vague notion hidden in the foreign section of newspapers. In China, only less than 20 million computers had access to the internet.

I wouldn't spare a thought for that odd moniker QQ for another two years, as I hunkered down for China's life-defining university entrance exam, sleeping four hours a day, blocking out everything and anything that didn't help advance my quiz skills. But as with millions across China, the name QQ gradually, almost insidiously, crept into my consciousness. And it was, in some ways, the only real thing that endured from two decades ago.

The telephone numbers my classmates scribbled hastily in my yearbook are no longer in use. Neither are the multitude of emails that ended in yahoo.com or hotmail.com. What remains is QQ and the company behind it: Tencent.

As a teenager, I had little concept of what life-long connections meant. Nor did I unduly ponder how technology morphs and shapes friendships and bonds. It's strange to think that the social networks, people and symbols we most associate ourselves with — the basis of our modern-day identities — are so tied up with big tech platforms. And how, when these companies grow outdated, parts of our lives vanish along with them.

Tencent's entire philosophy is to be a connector. It strives to link content, information and people, helping a billion users build their identities.

Its story is about the struggle to stay relevant, to avoid the fates of the many technology giants that have made way for newer models. This book is about that struggle, and the story of a company whose success eventually threatened its very existence, potentially upsetting the political overlords at home.

It's a story about a company that many outside China – not unlike much of my junior high class two decades ago – struggle to grasp, years after it overtook Facebook in 2017 to become the fifth largest company in the world.

Little known to people beyond the tech community, Tencent's sphere of influence extends far beyond its home turf. It reaches the screens of hundreds of millions of global gamers via titles like Fortnite and movie-goers via Hollywood blockbusters like Men in Black: International and Venom.

Backing some of the most popular global goods and services, including Tesla, Reddit, Snapchat and Spotify, it's the puppet master that merges the functions of WhatsApp, PayPal, Facebook, Uber, Deliveroo, Yahoo, TikTok into one super-app known as WeChat.

Tencent has been one the world's most powerful companies that few people outside its home turf are aware of.

That's changing. Amid escalating tensions between China and the US, global audiences are increasingly captivated by the power-brokers in the world's second-largest economy. From Huawei and Alibaba to artificial intelligence surveillance startups, China's tech giants have been thrust under the spotlight. Their fate amid a sweeping industry crackdown stemming from President Xi Jinping's campaign to 'curb the disorderly expansion of capital' is even more intriguing.

When I started writing this book, I was inspired by the notion of unearthing the story behind one of China's biggest entrepreneurial success stories. For years, I've heard people lament how China creates the best tech entrepreneurs but few writers bother to document their stories for a global audience – the discourse has swung from pandering obsequiousness to casual dismissal, based on the simple notion that their achievements are of no merit because everything is controlled by the government. The truth lies somewhere in between.

I wanted to write a book that would open doors to understanding the broader landscape of the rainmakers who uphold China's startup and venture capital maze, to decode the rising class of magnates underpinned by technology and capital, and more importantly show how technology services run by a presence halfway across the world could affect the lives of people living in Europe and the US. I wanted to document this era before our collective memories got distorted and we missed the opportunity to tell the stories of these entrepreneurial gladiators.

As I progressed on this humbling journey, the narrative for China's tech industry changed drastically. What we are now facing is a paradigm shift in regulation in China's political landscape – the government is hell-bent on reining in its biggest tech champions, which have amassed data from more than a billion users. For the Communist Party, ensuring stability at home – which entails keeping the rising class of wealthy, tech-savvy moguls in place before their economic aspirations turn political – is ever more pressing as tensions with the US brew.

The more important and interesting question now is: what does it take to be a successful entrepreneur in Xi Jinping's China?

For years, Tencent flew under the radar by design. Tencent's billionaire founder Ma Huateng, who goes by the English name 'Pony', shies away from media attention. His cordon of communications chieftains have swatted and stonewalled media queries with vehemence in the past. If his arch-nemesis, Alibaba founder Jack Ma (not related), is known for his high-profile, self-aggrandising personality, then Pony Ma has made a career of hiding behind the scenes.

Tracking Pony became something of a sport. It required tracing and chasing him down across cities at conferences and forums, interviewing his cotorie of confidants and conducted hundreds of interviews with Tencent's executives and staff at a conglomerate that spans gaming, payments and finance, cloud computing, news portals, instant messaging, video and film streaming and production. In the decade that I've been writing about China and its bustling internet industry, documenting the rise of Pony Ma has granted me a front-row seat to understanding the unpredictability and ferocity that defines China's startup landscape.

At its peak, half of China's ten richest people hailed from the tech scene. Alliances and allegiances change in the blink of an eye, national champions become public enemy No. 1 overnight. To get to the top, a generation of young entrepreneurs have had to cope with the vagaries of a suspicious, if not draconian, government, while spurring on an industry bloodbath by waging all-out war with thousands of headstrong entrepreneurs intent on building the same types of businesses.

It did seem, at one point, that Pony was ready to step out of his shell. In January 2015, when Tencent was on a tear – expanding in everything from finance to gaming and Hollywood and surpassing Amazon in valuation – I sat down with him at Tencent's headquarters in Hong Kong, along with a small group of select foreign-media reporters.

Slight and bespectacled, Ma walked into the meeting punctually, and sat across a boardroom table so wide, it seemed the sheer scale of it could help deflect unwanted interrogation. The self-conscious Chinese mogul had come a long way since his callow years as a programmer, swapping his unkempt mop of hair with a hard-part comb-over. The meeting was Pony Ma's way of sharing his ideas about what kept him up at night, where the juggernaut was headed, and to shed light on his empire's relationship with the Chinese government.

It was also his gambit for greater presence on the global stage. Whether it was whisking in soccer superstars like Lionel Messi to campaign for WeChat, or clandestine negotiations with Elon Musk and Evan Spiegel for stakes in their companies, Tencent one way or another has been able to insert its presence into the tentpoles that uphold Silicon Valley. 'Internationally, China has become a leader for quite a few mobile internet products and services,' Pony said[1].

At the time, China produced two billionaires every week. The world watched what once was dubbed the Middle Kingdom leap into the digital age in broad brush strokes. Behind the hyperbole of that internet explosion – a thirty-six-fold jump in online population; the birth of one unicorn, or a company worth more than $1 billion, every 3.8 days – there was also a Wall.

The Great Fire Wall of China: one out of three people using the web access it through a filter that obscures Facebook, Twitter, Snap, Instagram, the *New York Times* and YouTube. In a sense, it's a parallel universe, where nearly a billion people live and thrive – much to Westerners' surprise – on China's equivalent of such mainstays. There's Meituan for Deliveroo, Didi Chuxing for Uber, WeChat for WhatsApp and Facebook. The services often are even better in terms of convenience and design.

That Swiss Army knife of a super-app WeChat is the most deft at merging the functions of all those mentioned above. Domestically WeChat is known as Weixin, and the company has made it a point of emphasising that they operate as two apps within and outside of the mainland. China's deficit of privacy controls also means its companies and government have an edge when it comes to collecting the data that empowers the algorithms that screen, monitor, name-shame and, sometimes, imprison its citizens.

The dynamics between companies and the authorities are like no other. Way before China began its unprecedented crackdown on the internet sector, I sat down once with an official and talked about the

---

[1] 马化腾:互联网产业走出中国道路 期待更多众创空间落地浙江_科技_新华网

vicissitudes that startups and entrepreneurs endure. 'No matter what kind of hotshot you are, we will always have a way of showing you who's boss,' the person said, making an off-hand remark about Pony Ma. 'Don't think because you control a billion users and moved to Singapore or some overseas country that we can't do anything about you.'

The person then delved into how when regulators felt Tencent needed to be taught a lesson, they would step up censorship efforts, block or shut down web services till the company got the message. The tactics were not always conspicuous. Given WeChat's overseas ambitions, they would sometimes disrupt its service for overseas users, delaying messages or transactions for just half a minute. 'That small hold-up is more than enough to drive users crazy and make people ditch the app altogether,' the person said. 'That's how you show them some colour.'

The Wall no longer resides just within China. When Chinese people travel outside of the country, the Wall follows them via their telecom providers. A person using a China Mobile SIM card is barred from roaming on Google even after venturing beyond the country. Authoritarian nations in Africa, Southeast Asia and Russia see the appeal of the model. They too want to create their own intranet.

As the internet splits in two, aligning itself between the American and Chinese models, Tencent's story offers a window into an alternative vision of what the global online sphere could become. Silicon Valley's unfettered techno-utopia, once deemed the superior if not sole approach, is now under siege. US President Bill Clinton famously said in 2000 that controlling the internet would be like 'nailing jello to a wall', that liberty would spread by cell phone and modem, inviting everyone to 'imagine how much it could change China'. With the emergence of companies like Tencent, a swathe of the world is increasingly keen to adopt China's tightly controlled and heavily censored web. It's a stunning ideological coup for Beijing that would have been unthinkable less than a decade ago. In December 2018, Pony joined Chinese government officials in declaring their country's destiny to become an internet power.

Tencent's reputation is, if anything, controversial. Permeating daily life in China – more than two-thirds of Chinese people use it for everything from texting to shopping, flirting, dating, watching videos, playing games, and ordering food and taxis – Tencent's apps have become ideal tools for mass surveillance.

The Communist Party finds the idea of Tencent both appealing and daunting. The story behind how Ma has been able to manoeuvre under the peremptory and watchful ruling authorities is one of the most fascinating untold tales.

There's a Chinese saying 'Li yu tiao long men' – 'a carp leaping over the dragon's gate'. Legend has it that if it manages to swim upstream and vault an arch atop a waterfall on the Yellow River, it transforms into an Oriental dragon, a snake-like creature symbolising imperial power. The story of China's internet tycoons like Pony Ma for the past two decades is that of a generation of carp becoming dragons. The twist, though, is that these idealist geeks, who ventured out to change the world, are now shackled and have become part of a system that they wanted to change. Once self-made dragons have achieved the level of success they have in China, the more important question seems to be when and how do they bow out unscathed?

This book is written for anyone interested in these questions. It's a book about Tencent, but also its compatriots – many of which grew up under Tencent's wing – that together created what's now been branded as a golden era for China's internet industry. It's for anyone who is a competitor, user, investor, an aspiring entrepreneur in Southeast Asia, the US or UK, an avid China watcher, an admirer of the self-made billionaire or just a curious mind asking what it takes to succeed in the world's biggest internet market under an ever more suspicious and watchful regime – one that rivals and increasingly thinks of itself as superior to the US.

PART I

ORIGINS

CHAPTER 1

BOOK OF PONY

On a scorching September day, fourteen high-powered Tencent executives trudged through the far-flung western reaches of China's Gobi Desert. They were travelling in luxury – prearranged tents, standby medical support and trucked-in water for showers. Yet some complained that the twenty-six-kilometre journey was the most brutal experience they'd ever undertaken. By the end of the first day, a petulant few petitioned to spare their blistering feet and pack their bags early.

At the front of the group was Tencent's billionaire founder Ma Huateng, often referred to by others as Pony. The self-made engineer-turned-mogul had in past years mostly taken his closest lieutenants on annual retreats to cushy Japanese resorts or the chi-chi hotels of Silicon Valley. A two-day trek across the parched desert of Gansu – a land of endlessly marching sand dunes and fifth-century Buddhist grottos that delineates the ancient Silk Road – was a peculiar choice.

It just so happened that as they arrived at the campsite and got within network range, everyone's cell phones started beeping. The company's shares had just hit a record high, making Tencent the most valuable company in Asia.

There was no celebration. That night, as they huddled around a bonfire, Pony insisted on forging on. So, after a restless night battling howling winds and billowing sandstorms, the group collected themselves and headed out at 4 a.m. again.

Martin Lau, Pony's right-hand man, would later recall how the trip offered an unvarnished glimpse into his boss's psyche. Temperate, stoic and almost irritatingly self-aware, Pony is a quiet but doggedly persistent force who rallies people, many of whom he considers smarter by his own account, to undertake his vision.

Pony's story is a once-in-a-lifetime tale. The reticent programmer's worth at his peak was $34 billion, six times the lump sum of the compensation for the two hundred most highly paid US executives. His journey towards becoming one of the world's most powerful men traces China's own tumultuous transition from poor, walled-off nation to economic powerhouse, propelling one-seventh of humanity into the digital age.

LAND OF COMPROMISE

It's a story of conflicts and compromise. Pony is a master of creating products so convenient and intuitive that billions of users want to join his network. Yet in the back of everyone's minds is the knowledge that their every move, location and utterance is documented and potentially scrutinised, a fact they're increasingly and openly reminded of by the Chinese government itself.

Nowhere is this contradiction more apparent than at Tencent's headquarters, centred in the heartland of southern Shenzhen's high-tech district. Once a fishing village adjacent to the British colony of Hong Kong, Shenzhen became the first test zone for the country's reform and opening-up, just seven years after Pony was born. In 1979, paramount leader Deng Xiaoping circled a plot of land facing Hong Kong across a shallow strait and designated it the first industrial free-trade zone that could attract foreign investment and experiment with new – and decidedly capitalist – business models. The district, once an eighth the size of Central Park, has since expanded into the now-famous giant Shekou district, inspiring state-endorsed ballads such as 'The Story of Spring'. Pony witnessed first-hand the city's phenomenally rapid evolution from swamp-land to world's factory floor, then to a sprawling metropolis melding technology, finance and manufacturing. Today, government officials like to talk about how it rivals the US Bay Area in breadth and sweep.

When I arrived in Hong Kong a decade ago, the city basked in the glory of being one of the world's foremost financial centres. I also noted how its residents watched with horror, perhaps with a trace of awe, as its urban neighbour swelled in power and importance, chipping away at Hong Kong's influence. Ironically, Shenzhen was once roundly disparaged by Hong Kongers as a cheap weekend getaway for saunas, foot massages, manicures and karaoke, complete with a thriving sex industry. In 2018, it surpassed Hong Kong in gross domestic production for the first time. Now one of the richest cities in China, it doesn't want for any of the modern accoutrements of a global city: shiny skyscrapers, soaring highways, even robot fights. Its most famous corporate citizens include the controversial telecom giant Huawei, the world's premier drone producer DJI, and of course Tencent.

Pony spared no effort when it came to building his future command centre. First started in 2012, Tencent's newest office building took five years and more than half a billion dollars to construct. Ma handpicked NBBJ, the architect responsible for Amazon, Google and Samsung's iconic headquarters. But the billionaire wanted it to be more than a statement of financial largesse. With its twin gleaming towers of glass and steel, he turned the building into one of the world's biggest laboratories for new internet services and connected devices. It features holographic tour guides, conference rooms that adjust temperatures based on attendance, and alerts for the best parking spots before commuters arrive. 'In China today, there has never been smart architecture of this scale,' said Ivan Wan, Tencent's then chief architect who oversaw the project. 'Using our building as a massive testing field for the next generation of smart devices and technology is what makes this project iconic.'

What struck me was, within the halls of a building that served as a towering paean to futurism and commerce, the Communist Party's influence was omnipresent. In its open-space reading room, alongside books about the cosmos and the ancient Greek and Roman empires, Chinese

President Xi Jinping's book – tabulating his speeches and thoughts on how to govern – featured on the most prominent shelves.

The company extols the victories of the Chinese Communist Party even in its gym. Imprinted on its 300-metre running tracks are penguin-shaped QR codes. When scanned with smartphones, it brings up stories documenting battle victories during the 'Long March' – the Red Army's epic flight from the then-ruling Kuomintang army (later branded as Chairman Mao Zedong's genius gambit to preserve the forces of the Revolution). One penguin brought up a picture captioned 'Bloodbath of the Xiang Jiang River', and another 'The Witty Conquest of the Jinsha River'. 'Party officials love this when they see it,' a Tencent employee told me as she clapped her hands in a gesture imitating the cadres described.

Pony himself is a member of China's legislative council, which, along with the country's elite (mostly Party members) convenes once a year in Beijing to discuss the nation's agenda. It's one of the rare occasions when the mogul steps into the limelight and offers glimpses into his persona. That doesn't mean he likes it. A flustered Pony was observed in 2019 fleeing a blogger who beseeched him to relax internet censorship. 'Can you ask them to not keep deleting our accounts,' the person pleaded. Pony uttered not a word as he dashed away.

His talent for operating in such an environment, a gift for survival as a private entrepreneur in a state-dominant economy, partly stems from his upbringing. Ma was born on 29 October 1971 on China's southern island of Hainan. He began life in the midst of Mao's violent Cultural Revolution, during which tens of thousands were persecuted – at the hand of Red Guards, sometimes their neighbours, colleagues and even their own children. Until today, few speak of the atrocities – particularly those that they themselves committed. It scarred a generation of people who were taught the value of erring on the side of caution and vigilance when it came to matters of politics.

Pony's father, Ma Chenshu, was a native of Guangdong, a region known for industrious businessmen. A Party member, the elder Ma relocated to the balmy backwaters of Hainan after cadres were asked to help with economic development. He rose through the ranks of the local port authority, taking stints as an accountant and then a resource planner.

Decades later, the island would become famous for fostering – at the time – the boom to end all booms, when a generation of young people was consumed by avarice and the pursuit of asset wealth only to be crushed by the bursting of a massive property bubble.

The Mas resided on the island's western tip, an area inhabited by the Miao minority, known for their tattooed faces, silver accessories and vibrant clothing. At night, a young Pony gazed into a sky full of stars, piquing his interest in astronomy. He clearly clings to his memories of growing up there, as Tencent would later choose Hainan as a key base for cloud-computing and a data centre.

Pony was thirteen when his father was reassigned to a Shenzhen state-owned port company. Leveraging his knowledge in finance and accounting, he became a manager, then board director of the company, giving his son an invaluable introduction to the intricacies of the corporate world.

PI DIGITS

Second child to an older sister, Pony was quiet, well-behaved and largely unnoticed during his school years, according to his teachers. At thirteen, he was a scrawny figure of little less than five feet, always sitting in the front row of the classroom. One thing that stood out was an obsession for telescopes. As a child, he crafted one based on instructions from a popular kids' science magazine. Pony says he still remembers that, in April of 1986, he was the first in his school to spot Halley's Comet thanks to a high-end 80-millimetre telescope that cost 700 yuan, four months of his father's salary, which he got for his fourteenth birthday. It didn't come easily. His initial pleas were dismissed. But his mum spotted in his backpack one day a journal in which the teenage boy declared that his dream of becoming a scientist had been squelched by his parents. His father eventually gave in and bought him the telescope. Pony used it to snap a photo of the famous comet, which he included in an essay that he mailed to Beijing as an entry in a science competition. He came in third. After Pony grew fully immersed in the digital world, he would seek solace in his childhood obsession. 'Doesn't the internet look like an uncertain, forever exploding galaxy?' he once said[1].

It was in middle school that he would meet three of his co-founders for Tencent: Xu Chenye, Chen Yidan and Tony Zhang Zhidong. All of them attended prep classes for the Math Olympiad, an international competition extremely popular among Chinese students. Only Pony joined an astronomy club. The teenagers would often spend time competing to recite Pi digits in the hallway after class. At one point, Chen recalls, they got to about a hundred digits.

Pony and his friends' obsession with STEM – science, technology, engineering and maths – was partly rooted in educational policies geared towards nurturing the talent needed to jump-start the country's industrial manufacturing apparatus, an attempt to get the economy back on track after decades of turmoil. For the first time in a century, China seemed to be on the brink of actual lift-off. The winds of change were stirring. In Pony's school, people repeated epigrams such as 'shi bu wo dai', meaning, time will not wait for me.

Then something happened just a year before Pony attended university in 1989. A rift in visions for economic development – state-driven versus market reforms – led to factional conflict within the ruling Party that eventually spilled into the public domain. Thousands of students took to the streets of Beijing, occupying the capital's iconic Tiananmen Square, demanding democracy and the freedom to write their own destinies. It so happened that Mikhail Gorbachev, General Secretary of the Communist Party of the Soviet Union, visited the city around that time – the first head of state to travel to China since the two countries severed ties more than three decades earlier. The foreign press covering that summit were captivated by the protests that subsequently erupted into a global spectacle. 'Unbelievable, we all came here to cover a summit, and we walked into a revolution,' a CNN correspondent in Beijing said at the time.

After weeks of agonising debates, a purge of the Party's liberal faction ensued, and the government rolled in the tanks to crush the movement. That period in history has now been wiped from Chinese textbooks and

---

[1] W, Xiaobo. Tencent. 1st edition, Zhejiang University Press, 2017.

the domestic internet sphere. It also templated the formula for dealing with dissent ever since – to maintain the one-man rule that defines the Leninist state that China is, only absolute loyalty to the paramount leader is of importance. It taught the Party to always keep tabs on public sentiment so as to nip protests in the bud. And no surveillance vehicle has been as powerful and efficient as the country's now ubiquitous internet service providers.

It was because of the turmoil that many parents, including Pony's, decided to let their children stay put in Shenzhen. He scored 739 out of 900 in his university entrance exam, enough to get him into Tsinghua or Peking University, the highest echelon of the nation's education arena. Instead, he enrolled in local Shenzhen University, a school that was only five years old and so nascent that it was encircled by farmland. There was no astronomy major, so the inveterate star-gazer opted for computer science. 'Pony Ma's class was the best batch of students ever,' Hu Qingbin, Pony's coding teacher, recalled. 'No one failed in class. This never happened before or after their year.'[2]

Pony was known among his college classmates as somewhat of a geek, an epithet that followed him through life. 'I was trained as a software engineer. I'm an entrepreneur, but I'm not eloquent because of my background,' Pony once said[3]. His middle-school classmate Xu Chenye also attended the same university. The two often jogged in the mornings and competed at memorising English words. At school, the usually quiet and diffident Pony developed a maverick streak, periodically hacking into university computers. When he locked the hard drives, even the administrators weren't able to free them. He rapidly earned a reputation as a computer maven. His talent was unmistakable: for his undergrad thesis, he created a software program to predict share price movements. Parts of it even touched on using artificial intelligence to forecast stock trends, a cutting-edge idea for his time.

Pony was twenty-two when he scored his first entrepreneurial coup, selling his stock-trading program to the company he was interning at. Cautiously optimistic, he asked for 50,000 yuan, about three years' salary for a fresh graduate. To his surprise, the company agreed on the spot.

---

[2] W, Xiaobo. Tencent. 1st edition, Zhejiang University Press, 2017
[3] HKU media files, June 2015

That's how he knew he was on to something big. Halfway across the world, a tech boom was in the making. The then-revolutionary Mosaic web browser came into being the year Pony graduated from college, making it possible finally for the masses to access the internet. HTTP was the new-fangled tech that triggered a spurt of growth, and Silicon Valley was wrapped in euphoric exuberance.

Pony's first job-hunt attempt went surprisingly well. He bumped into an acquaintance in a bookstore who told him a pager-maker called Runxun was hiring. He went in for an interview, touted his experience writing software and got an offer the very next day. For the benefit of those born in the smartphone era, pagers were a small device people used to clip to their belts, which vibrated or beeped to tell you someone wanted you to call them. It was a status symbol, the 'it' gadget back then. The budding entrepreneur witnessed his employer's boom then bust in just the five years he worked there. His salary jumped eight-fold, but he grew dissatisfied being one of myriad dime-a-dozen middle managers.

It was in the virtual world that he found his voice. In 1994, Pony grew captivated by something called FidoNet. This was a network that enabled emails and file-sharing between bulletin boards, themselves a popular form of communication before the advent of low-cost internet connections. Pony was intrigued by how one could meet people from all corners of the country who share a common interest. 'I felt like it opened up a new door, it was the start of the internet,' he said years later.[4] He set up his own bulletin board, naming it ponysoft. Unlike the internet, FidoNet allowed just one user per telephone line; so he set up four phones and eight computers at home, costing him nearly 50,000 yuan. It meant spending almost all of the money he got from selling his software. That's a pattern Pony clung to throughout his life: when he has his heart set on something, he doesn't hesitate to invest.

The reticent geek became a completely different person in the virtual world. 'He would just prattle on and on,' obsessing over technical challenges and responding to every issue raised by visitors to his bulletin board, fellow netizen Li Zonghua said.[5]

[4] 吴晓波. 腾讯传1998-2016:中国互联网公司进化论. 浙江大学出版社; 第1版.

[5] W, Xiaobo. Tencent. 1st edition, Zhejiang University Press, 2017

In fact, Pony was to become so comfortable in the digital world that he was later to find his future wife there. He would go on to chat anonymously with a music teacher called Wang Danting for three months, before meeting her in person. Love swiftly blossomed and the two would go on to marry six months later. But all of this wouldn't be possible if Pony and his team hadn't created their first hit internet product: QQ.

CAMBRIAN EXPLOSION

In April 1994, China and the US for the first time agreed to link up their internet. The National Computing and Networking Facility of China, a state-backed project, leased a dedicated 64K line through Sprint Co. China was officially, now, part of the world wide web.

It ignited a Cambrian explosion of the Chinese online space. Jack Ma, a thirty-one-year-old university English teacher, set up his first website in May 1995: an online yellow pages that allowed companies to post information online. Charles Zhang Chaoyang raised $1 million from US investors, ditched his dream of becoming a physicist, returned to Beijing and eventually founded one of China's largest online portals, Sohu.

The tidal wave of change made Pony restless. While most people had never heard of, let alone comprehended, the concept of a virtual world, he had already been living it. Seeing the opportunities to come, he took action. That's when Tony Zhang Zhidong came back into the picture in almost theatrical fashion. The two classmates had lost touch after college and Zhang was working at a rival pager company. At the time, he was looking into a malfunctioning server that he suspected was the victim of a malware attack. Zhang traced it back to Pony's employer, Runxun. He knew of only one person who was capable of such an act; he picked up the phone and called Pony, 'It was you, wasn't it?' Pony chuckled in response. 'I was only trying to test your skills.'[6]

---

[6] W, Xiaobo. Tencent. 1st edition, Zhejiang University Press, 2017

After the Lunar New Year of 1998, Pony finally reached out to Zhang for coffee. He was ready to build a startup, and he'd picked his old friend to do it with.

Back in the real world, Pony didn't have a clue what his startup's business model would be. The rough idea was that they would create a product that combines the pager and the then-nascent internet. Pony reconnected with two other classmates: Chen Yidan, who was working at the Shenzhen quarantine bureau, and Xu Chenye, who was part of the telecommunications bureau. They had one big problem: none of them knew anything about sales.

Enter Jason Zeng Liqing. Unlike the initial founding quartet of self-proclaimed nerds, Jason was an outgoing, articulate and towering presence. A cadre at the Shenzhen telecommunications bureau, he once convinced a local property developer to invest 1.2 million yuan in building the first walled-off compound in the country to be entirely covered by broadband. The five clicked, and Jason took on responsibility for sales. Pony would take charge of product and strategy. That division of duties between the founders, and Jason's own dominant personality, planted the seeds of a rift that would eventually see Jason marginalised.

Pony registered a startup for 500,000 yuan, equivalent to sixty-two years of the average Chinese wage at the time. His dad helped him file for the company name. After trying out three that were already taken, he requested the moniker Teng Xun – the first character alluding to part of Pony's Chinese name (Ma Huateng), the second meaning speed and information. In English, the name Tencent paid homage to Lucent Technologies. Because both Pony and Zhang Zhidong hadn't officially resigned at the time, they put Pony's mother down as owner – for about a year, the Chairman of Tencent was his mum, Huang Huiqing.

DISCO BALL

Looking back, the two years through 1999 were a watershed moment in global internet history. That was the year Jack Ma started Alibaba, which would go on to become the world's No. 2 e-commerce company. Steve

Jobs introduced the iMac. Microsoft bundled its software into its Windows operating system, triggering a US Justice Department antitrust suit. Yahoo's Jerry Yang made it onto the cover of *Time* – and made one of the worst mistakes in his life when he spurned two Stanford graduates who tried to sell him their search technology for $1 million. That turned out to be Larry Page and Sergei Brin, who founded Google. Missing the internet wave during those years meant omission from an entire era.

Tencent's early days were far from glamorous. Pony and several other co-founders set up shop in what has now become something of a domestic tech-industry landmark, the storied neighbourhood of Hua Qiang Be: a raucous electronics bazaar spanning several blocks, where thousands of vendors hawk everything from knockoff phones and laptops to sophisticated surveillance cameras. Today the government has given its main commercial drag a makeover, delineating it with gleaming skyscrapers. Peek inside, though, and one realises little has changed. Young boys and girls, many looking no more than seventeen, toil in dimly lit malls behind cramped stalls, slapping circuits on boards and wrapping packages for shipment as impatient delivery men look on. It's a cacophonous symphony of unwinding duct-tape and uproarious 1990s disco music.

The new company's office was a crammed 30-square-metre room with a disco ball dangling from the ceiling. Slot in a few chairs and desks and it felt jam-packed. Pony's idea was to create a product based on the beeper business. The device was considered a status symbol, hence high-margin. But linking the company's future to a hardware device confined growth. Devices weren't instantly scalable like internet platforms. It was also extremely competitive. It trapped them in a rapidly obsolescent model. No one wanted to use their software either, so Pony cut the price tag on his internet-connected beeper service to a seventh of what he originally sought, barely breaking even. To generate extra revenue, they moonlit, developing software for an email system. Pony didn't dare ask two of his co-founders to quit their full-time jobs, for fear of screwing up their lives. Tencent teetered on the brink of collapse just about every single day.

It was in a fit of desperation that Pony stumbled on a service that would save the company: ICQ, a ground-breaking online chat system created by five Israelis in June of 1996. At the time, the non-tech-savvy were using computers powered by non-Unix systems like Microsoft Windows.

Derived from the English phrase 'I seek you', the free-to-download product was built on the concept of a fully centralised service with individual user accounts focused on one-on-one conversations. ICQ became the first widely adopted instant-messaging platform. The huge success of the online chat system caught the attention of AOL, which bought the startup behind ICQ in 1998 in its largest-ever acquisition.

Pony was wandering the streets of Guangzhou in August 1998 when he spotted an ad posted by the local unit of China Telecom, seeking bids for developing a software system similar to ICQ. The co-founders at Tencent thought it was technically feasible but, to their disappointment, the state-owned company had already picked a supplier. A huge debate ensued on whether they should pursue a parallel project. The main point of contention was how to make money off it. The answer was no way. But Pony prevailed over his colleagues. 'How about we nurture it first,' he suggested.

Thus began the story of Tencent's controversial origins – being a copycat. Just as many of China's largest internet companies started out by imitating Western peers – Sohu from Yahoo, Baidu from Google, Weibo from Twitter, Alibaba from eBay – Tencent's first hit came from emulating ICQ. In a sense it set the tone for the company's later practice of replicating others – something that would come back to haunt Pony.

Pony's own initial vision for his chat software OICQ was timid, to put it mildly. The subsequent explosion in user growth and his own staff surpassed his wildest dreams. Three months after founding Tencent, the co-founders launched the desktop chat service. Registered users were assigned a series of numbers for account names. Pony reserved the first username – 10001 – for himself, and kept the rest of the first two hundred for employees.

The co-founders thought OICQ would be considered a hit if it lured ten thousand users in three years. To maintain service, they conservatively estimated the company would only need about 100,000 yuan a year. Indeed, things started out slowly. 'When users first came, nobody was in the chat room. I had to keep them company and talk with them,' Pony recalled years later. 'Sometimes I had to change my profile picture and pretend to be a girl.'[7]

---

[7] HKU media file, June 2015

CRASHING SERVERS

Two months after launching OICQ, an exhausted Pony and his co-founder Chen Yidan logged on one day after a fruitless day of trying to sell their beeper software. They were thrilled to discover that the little online message service they regarded as a sideline had hit five hundred users. They celebrated this victory with a bottle of cheap beer.

The founding team had one thing going for them: they understood Chinese users. In 1998, less than 1 per cent of the population had computers. Most people accessed the web from so-called internet cafés – smoke-filled rooms with rows of desktops, brimming with bloodshot-eyed teenagers who often spent too many sleepless nights playing games.

So Tencent made micro-innovations to the service that increased its appeal to those who didn't own computers. Unlike competitors that only allowed users to store contact lists on their desktops, Tencent's users could see their social network on any computer because it kept that data on its servers. Pony also made its chat software easier to download. With Chinese internet speeds at an average of 28 kilobits per second, most rival products were too big and took more than half an hour to download. Tencent was able to shrink OICQ to less than a tenth of its competitors' size, taking only five minutes to install. And unlike ICQ, Tencent let users send messages to friends who were offline, and add strangers who were online as well. These two features laid the foundation for its eventual popularity as a social networking tool.

Those long, sleepless nights paid off. After the company fixed bugs on a daily basis and rolled out updates as frequently as three times a week, OICQ users began to spike as much as four times weekly. It grew so rapidly that Tencent's servers were constantly on the verge of crashing.

Unlike in the US, servers were expensive in China. Tencent chief technology officer Tony Zhang almost literally beat his head against the wall finding ways to improve his algorithms and lift the computing burden off CPUs. That ingrained habit of operating off a shoestring budget helped Tencent garner much-needed tech firepower and capability in its early days that would help in later arms-races with rivals.

OICQ's preliminary success also brought about other headaches. In early 2000, AOL (American Online), which bought the Israeli instant-messaging company ICQ, filed a lawsuit with the National Arbitration Forum in the US, accusing Tencent's OICQ's domain names OICQ.com and OICQ.net of violating intellectual property rights. Tencent lost the case and had to shut down OICQ's websites. Tencent.com became the main portal for users and Pony changed his own chat service's name to QQ instead in December that year. Tencent also voluntarily removed avatar images that were Disney cartoon characters such as Mickey Mouse and Donald Duck.

During those days, Pony was fraught with anxiety. A year into his startup, OICQ still wasn't making money. His team needed to take on sideline assignments just to make a quick buck and sustain the business. Their service's signature message alert – a beeping chirp resembling that of a pager's – began to haunt the co-founders. 'It was like a pixie who was a ghoul in a past life, always hungry,' said co-founder Xu Chenye.[8]

In a last-ditch effort, Pony and his team made an appeal for IDG Capital to invest. Backed by the late Patrick McGovern, IDG was one of the earliest global venture capital firms to set foot in China.

'Pony did the presentation. He was intellectual, and extremely shy,' Hugo Shong, a founding partner at the investment house, told me years later. Shong translated for Pony because his English wasn't that great back then. Even today, the billionaire still prefers to speak in Chinese during forums and interviews.

Though Shong had his doubts, IDG joined PCCW, the Hong Kong conglomerate backed by Asia's richest man at the time, Li Kashing, in staking $2.2 million for a 40 per cent slice of Tencent in April 2000. 'We thought they learned fast and there was good personal chemistry.'

With the freshly injected capital, Tencent lived to fight another day. But not for long. The tenacious founder would soon once again find his proud creation almost bankrupting the company.

---

[8] W, Xiaobo. Tencent. 1st edition, Zhejiang University Press, 2017

CHAPTER 2

MARCH OF THE PENGUINS

Hadspen House is one of the most beautiful historic estates in England. Its private trails meander through 300 acres of woods and gardens graced with fountains. The man who bought the property for 13 million pounds in 2013 was Koos Bekker, the Chairman of the South African telecommunication company Naspers Ltd, a South African Apartheid-era linchpin turned modern-day technology investment giant.

The wealth amassed by Bekker hails predominantly from China, and more specifically Tencent. Much to Naspers' fortune, and perhaps chagrin, at its height its stake in Tencent was worth nearly three times the entirety of its own market value. The story of how a South African company came to become the biggest investor in China's national champion started in one of the smoked-filled internet cafés mentioned earlier.

Around 2000, the internet bubble burst and Silicon Valley was in disarray. Microsoft plummeted 63 per cent from its peak, Cisco shed eighty per cent of its value and never managed to climb back to its 2000 heights. The panic rippling across the US spilled into the China tech space. That meant trouble for Tencent.

Despite explosive user growth – close to hitting the 100 million benchmark – Tencent had yet to figure out a business model. It reached out to its two backers for money multiple times, to no avail. IDG Capital, sensing pressure from the US, decided to cash out. With the market in free-fall, no one was interested in buying a cash-burning operation, despite it being one of China's fastest-growing companies. The Tencent founders considered selling out to the leading internet portals in China, such as Yahoo, Sohu and Sina. They were flatly rejected.

At its bleakest moment, an American by the name of David Wallerstein showed up at Tencent's office in Shenzhen. He said he represented the largest telecom company in South Africa, MIH (Naspers Ltd). Wallerstein spoke fluent Chinese and had seen young people everywhere in China using Tencent's chat software.

He was an odd one out. Back then, few foreigners ventured outside the larger cities such as Beijing and Shanghai. Tencent's headquarters in Shenzhen was for the more daring drifters. His curiosity to explore stemmed from a wandering childhood. Wallerstein attended high school in Japan, and by the age of sixteen he learned to be at ease as the only non-Asian in a room. Thinking China was destined to become a superpower, he went there to work as a management consultant, helping foreign companies break into the market. His clients were predominantly in the IT and telecommunications industry.

At twenty-one, he came across the opportunity of his life. Naspers was on the hunt for investments, following a windfall after it went public on the Johannesburg Stock Exchange. Wallerstein parleyed a role for himself within the South African company to scout for companies. His strategy was going local.

'Whenever I visit a Chinese city, I visit the internet cafés to see what games the young people are playing. To my surprise, almost all of the internet cafés had OICQ on their desktop. At the time, I thought, this could be a great internet company,' he said.[1]

He offered Pony a valuation of $60 million, requesting that Naspers become the largest shareholder in Tencent and pay for the investment with Naspers shares. To his surprise, Pony and the founders heard him

---

[1] Y Combinator interview

out but politely refused. For Pony, it was essential that the team keep control of their company.

In private, Pony felt a reason to be happy for the first time in months – someone at least was willing to invest in his baby. Wallerstein had offered a price almost double that of what Tencent was worth a year ago.

Wallerstein wasn't about to give up without a fight. He managed to eventually convince the young founders to dinner. Over a boozy meal, he got them to warm up to him and followed up by going back to their office the next morning. This time he brought forth proposals on how Naspers could help Tencent import ideas or technology from the US. That garnered Pony's interest. 'Ultimately I felt like we needed to demonstrate we could provide some unique value,' Wallerstein said.[2]

It took him months of negotiations and lobbying at headquarters, but he convinced Naspers to let the Tencent founders maintain control. Naspers also paid in cash. It offered $32 million to become the largest external shareholder in Tencent in 2001. IDG kept 7.2 per cent of its stake after selling 12.8 per cent, making an eleven-times return. PCCW sold all of its 20 per cent stake. To this day, Naspers remains Tencent's biggest stakeholder.

As part of the investment, Wallerstein decided to move to the US immediately to scout for opportunities that could benefit Tencent. He'd travel back to Shenzhen every other week to catch up with the founders. The model worked so well that Wallerstein joined Tencent in 2001 when the company only had forty-five people.[3]

BREAKING EVEN

With a freshly replenished war chest, Tencent was bent on figuring out a business model. The answer came in text messages. In the US, SMSs

---

[2] Y Combinator interview
[3] When I met up with Wallerstein in 2017, he was working for Tencent out of California as its chief exploration officer, looking for moonshot projects, including Moon Express, a startup that aims to put drones on the lunar body; Argentina's Satellogic, which specialises in satellite imagery; and Planetary Resources, which is looking into asteroid-mining.

hardly took off until iMessage emerged many years later. They were costly, cumbersome and counter-intuitive. Users were limited to a certain number of texts with their mobile plans. In China, phone calls were significantly more expensive than texts. Chinese users sent nearly the same number of messages in a day on Lunar New Year in 2001 as the US did in an entire year.

The following year, China generated a third of global SMSs. Inspired by Japan's NTT Docomo, Pony proposed to one of the country's largest telecom operators that they ask users of QQ to register for a mobile account via text, and get alerts via QQ's desktop software as well. People paid only 60 cents a month to get alerts on their mobile phones for messages they received on their QQ desktops. While Tencent didn't charge directly, the state-owned China Telecom would split the revenue generated from add-on value like images and music. Tencent would funnel its traffic and help the telecom operator generate more customers. The company broke even for the first time.

It was a major coup. Ironically, that placed a lot of pressure on the QQ team. The desktop instant-messaging software itself was not generating money. Pony's first few attempts at direct monetisation were not successful. Among them: placing banner ads on QQ, pushing for membership fees, selling QQ's service to companies – a sort of 1.0 version of Slack, an office messaging tool. In hindsight, those concepts were great, but too far ahead of their times. For startups, that's worse than being late to the game.

QQ worked like a contact list and msg service. When people were offline, their headshot profiles showed up as black and white, but they turned colourful when they reappeared online. Tencent allowed people to add a few words to depict their mood or status, offering some form of differentiation. That became a primitive social network.

Users were scaling exponentially, but they still refused to pay, part of the culture that dominated China's internet back then. In order to control cost growth for servers, Tencent started imposing quotas on registration, slashing them by 60 per cent from 1 million per day. That prompted people to sign up multiple times, pushing success rates to as low as 2 per cent. Tencent tried to encourage everyone to apply for accounts by texting, which cost a fee. The public, accustomed to free services online, was put off by the idea, growing enraged by the difficulty and money it required to get a QQ account.

Pony was unperturbed. He insisted during internal meetings on keeping the fee-based model. In those early days, making money was clearly more important for survival than being popular. 'Tencent won't die because people are scolding us, but if we failed to find a profit model we would go belly-up,' Pony said.[4]

The founder ended up learning a lesson. Competitors sensing the opportunity to lure QQ users rushed to offer the same services. Yahoo, Skype and a slew of local competitors scurried to create their own instant-messaging platforms. All of a sudden, Tencent had lost its monopoly and was fending off more than thirty rivals.

On QQ's third birthday in 2003, Pony decided to offer QQ up for free again, granting each user a long-term account without charge if they registered via mobile text. Not long after, it completely stopped charging people. Pony learned quickly that the notion of being on top is hard to sustain, especially since the country was notorious for lax protection of intellectual property rights. Any idea or design could be replicated. The hounds were constantly around the corner waiting to rip off any successful product. In fact, Tencent itself would become the most infamous copycat.

With QQ as a free product, Pony was on the search for a viable business model, again. Internally, Pony's team had come up with the idea of selling virtual coins. They just needed to find a reason through which to lure people to buy them.

In South Korea, sayclub.com, a virtual community, was experimenting with the concept of avatars. They ranged from designs as simple as a cartoon icon to the more sophisticated ones that allowed people to buy clothes or special characters to demonstrate their personalities in the online world. It quickly became a pervasive feature for the top Korean site. Pony's team, not

---

[4] W, Xiaobo. Tencent. 1st edition, Zhejiang University Press, 2017

understanding Korean, spent 400 yuan asking people to translate sayclub's content. They were still late to the game, beaten by Chinese rival NetEase.

Tencent had to speed up. Pony made the decision to set up the 'Avatar Team' to create eight hundred virtual designs – clothes, necklaces, sunglasses, suits, different hairstyles – that became known as the QQ Show. It went live in January 2003, with the company giving away ten virtual Q coins to its first batch of users.

The product became most popular with young people in their early twenties. The basic props weren't expensive, costing as little as 10 cents. Yet to establish their roles in the world, users eagerly sought to differentiate themselves via their online personas, which included spending bigger dollars to dress up their avatars. Many people were using QQ as a dating app, adding strangers randomly and chatting with them behind their veils. It was important for some to establish themselves as high rollers.

The online world also offered an outlet for Pony, who was among the first batch of users of QQ Show. The avatar he bought had long hair, sunglasses and skinny jeans, looking very much like a cowboy, a far cry from his geeky persona in real life.

'Tencent wasn't selling clothes, it was selling emotional support,' Pony said at the time. 'When friends see what I'm like on QQ, they will understand what kind of person I am.'[5]

For the first time in a long while, Pony felt QQ was in a good spot. He began thinking about what other services he could add so that users would stay on the platform.

QQ NEWS

Pony thought the time was ripe for Tencent to expand into the news portal business. He'd been keen on the idea of building an aggregator for information – as Yahoo had done – for quite some time, but was vetoed internally by his co-founders several times.

---

[5] W, Xiaobo. Tencent. 1st edition, Zhejiang University Press, 2017

The concept of a news portal was nothing new in China. Competitor Sina had set up such operations since as early as 1996, followed by Sohu and NetEase. It was a competitive landscape due to the promising revenue prospects, especially given the success of Yahoo in the US.

Daniel Mao, the son-in-law of Hu Jintao, who later became Chinese president, had been working for Sina as chief operating officer since 1996, underscoring the growing prestige of the internet sector. When Sina went public in 2000, Mao ranked eleventh on a list of China's richest IT entrepreneurs published by Euromoney Institutional Investor, which estimated that his personal wealth stood at $35 million to $60 million.

Pony argued that it strategically made sense to build Tencent's own news portal. His plan was to revamp an existing online community that had about 1 million users operated by Tencent and incorporate news content produced by other media outlets. They poached a senior executive, Sun Zhonghuai, who had worked at NetEase, allowing him to set up base in the capital of Beijing. The company also got lucky when it came to finding the right IP address for its new endeavour. Pony had had his eyes on the domain www.QQ.com for some time. The owner previously demanded 20 million yuan, way out of reach for the company. Luckily, when Wallerstein reached out in his own capacity again that year, the owner agreed to sell it for just 0.3 per cent of the original price. After three months of internal testing, the service went public.

The portal fit Tencent's vision of integrating different products – linking instant messaging with social media, music and news. One design idea was letting QQ users get pop-up alerts from Tencent's news portal on their instant messaging software when breaking news happened.

'We started building an intricate web of experience services that our competitors were just not doing,' said Wallerstein. 'It was very valuable in differentiating ourselves from other instant-messaging companies,' making QQ's business very defensible.[6]

It was in those early days that Pony started presiding over bi-weekly meetings with the company's core management team of a dozen people.

---

[6] Y Combinator interview

Starting at 10 a.m. every day, the meetings usually lasted well beyond 2 a.m.

Pony liked to listen to everyone's opinions before the group made a final decision. Project managers often played an important role in decision-making. He rarely used his veto rights, acting more like a mediator. Some of the most important decisions Tencent made – including the green-lighting of its signature product WeChat – have been struck after midnight. When proposals receive a majority vote, they get passed. The marathon-like meetings were a stress test on the executives physically and mentally.

GOING PUBLIC

In 2003 Pony met Martin Lau, the bespectacled and measured Goldman Sachs banker who would become his right-hand man.

Pony was reserved at their first meeting. 'He doesn't do chit-chat, he doesn't do small talk,' said Martin, recalling that Pony only became talkative when he went over his product.

The two met in the lobby of the Regent Hotel in Hong Kong to explore the possibility of an initial public offering (IPO). Tencent was one of the first listings of a Chinese internet company since the dot.com bubble. Martin specialised in telecommunications and media for Goldman's Hong Kong office at the time.

In the course of angling to handle the IPO, he and his colleagues scrambled to appear savvy, even skirting their firm's network firewall by asking a colleague in Beijing to sign them up for QQ accounts. Then they printed their usernames on new business cards before meeting up with Tencent.

Martin further impressed Pony with his upbringing. He spoke fluent Mandarin, something of a rarity back then for Hong Kong bankers, who mostly conversed in English and Cantonese. Born in Beijing, Martin's father was Indian-born Chinese, and his mother Indonesian-born. Both of Martin's parents were electronic engineering majors who responded to

Mao Zedong's call to the diaspora to help build the motherland after the Communist Party took over in 1949. They met and married in Beijing.

Martin and his older sister were born during the final years of the Cultural Revolution. The family left for Hong Kong when he was six, during the reformist regime of Deng Xiaoping. 'History happened and eventually they felt they needed to find a better future for their kids,' said Martin. 'It was a painful period.'

The family originally planned to move to Pakistan, where his grandmother resided. Yet en route they landed in Hong Kong, a bustling city that was fast becoming a trade and commerce hub. They settled in what was back then still a British colony.

Martin, whose full Chinese name is Lau Chi Ping, spent the rest of his childhood in Hong Kong, playing video games on his family's Apple IIe and dreaming about building rockets. His parents, he says, 'gave us this notion that if you're an engineer, you can always survive in a lot of turbulence, because your skill is always needed'.

He tested at the top of his prep school class at King's College, then, after concluding that only Americans could join the space race, decided instead to study electrical engineering at the University of Michigan in Ann Arbor.

He was picked up at Detroit Metropolitan Airport by Arthur Yeung, a professor of management who would become his mentor and eventually a Tencent adviser. Yeung was also from Hong Kong, and he often helped his countrymen get situated. He introduced Martin to a church near campus. Martin started attending regularly; there he met his future wife, Millie, a business-school student also from Hong Kong.

In the spirit of all overachievers, Martin loaded up on academic credentials, graduating from Michigan and then getting an engineering master's from Stanford and an MBA from the Kellogg School of Management at Northwestern University. After school, he went to work at McKinsey & Co., then Goldman.

He and his team bolstered their chances at winning a mandate to work on Tencent's IPO by studying up on Tencent. Martin, not a natural QQ user, marshalled his team on the ground in China to study how young people – especially those twenty-five and below – were using the software that had become pervasive in Chinese internet cafés.

Martin was forthright. He told Pony that Tencent's revenue model at the time was solely reliant on Chinese telecom carriers – users who paid would receive texts on their mobile phones when their friends sent them messages via the desktop QQ software – making the company vulnerable. He suggested Tencent focus on the network effect of QQ, its potential to cross-sell other services like news, which would come across as more appealing to investors. Pony was titillated by the idea.

Pony felt like he'd found someone who clicked with him. Earlier, he had brushed off rejections from his co-founders and team, insisting on expanding into games and a news portal precisely because he harboured the same concerns.

To add more value, Martin arranged role-playing sessions to show Pony the ropes – how to lure capital. They invited an investor to pretend to be a demanding backer, while other bankers played the co-founders. Martin's boss took on the role of Pony. 'We built a strong image,' said Martin.

Goldman won the mandate in September 2003. Martin worked on the IPO for nine months, pulling off eighty meetings with investors, including in the US and South Africa. One of the biggest debates at the time was whether to list the company in New York or Hong Kong. Choosing the latter would have been a rare thing for tech companies at the time. That sparked intense debates among the founders.

Most of the first-generation internet companies coveted a listing on the tech-heavy Nasdaq or the New York Stock Exchange. 'We felt that eventually the China market was going to be big, and we wanted to stay close to our end users,' said Martin. In the end, Pony took Goldman's suggestion to list in Hong Kong.

As one of the first tech IPOs in the city, Martin and Tencent received a ton of questions from the stock exchange. In Pony's discussion with Martin, the founder was okay with the fact that Tencent wouldn't be fetching valuations as high as in the US.

'It was the right thing to do instead of having the wealth pop,' said Martin. 'There would be a lot of disruption you bring to investors, if the expectations are very high; you bring disruption to employees, because they get the shares and all of a sudden they become incredibly rich.

'It's not very good for you to continue to hire and reward people. It's not very good when you continue to issue stock, it's much better when the growth is slower. That matched what the founders wanted.'

During the IPO roadshow, Chief Technology Officer Tony Zhang and co-founder Charles Chen turned to Martin out of the blue on their flight, asking him if he would consider joining. Pony and his co-founders had computer-science backgrounds, but they had little international experience and knew they needed help building a sustainable business.

As their banker, Martin thought it was a conflict of interest. But he also admits that he wasn't entirely sold on the job's prospects. 'To a certain extent, I was trying to see if this was real or not,' he says.

The company went public in June 2004, raising HK$1.4 billion ($180 million). Its early expansion into other sectors like games showed promise, boosting revenue 55 per cent that year. By the end of 2004, Pony and his co-founders, by then multimillionaires, once again offered Martin a job. This time he said yes.

Tencent's second offer resonated with him, he says, because it would allow him to tap his engineering skills. He accepted – taking, he says, a significant pay cut. Pony initially gave Martin the title of chief strategy officer and placed him in charge of investor relations and mergers and acquisitions, two fields largely unknown at the time in China.

Martin brought standard US corporate practices, such as setting revenue goals, to Tencent, and developed a five-year plan to enter new businesses such as social networking and digital media. 'This was a discipline that was urgently needed for a young company growing extremely fast in 2004,' says Hans Tung, managing partner at GGV Capital and a co-investor with Tencent in Didi, the Chinese ride-sharing company.

TELECOM SETBACK

Tencent had another reason to be happy that year. Pony, who turned thirty-three that year, got married. His wife, Wang Danting, met him online via QQ, originally oblivious to the fact that he was the founder. A popular version of the story among staff is that Pony had temporarily resumed authorisation for friend requests so he could add a colleague. While he was waiting, he received a message from a stranger who asked him, 'Who are you?' He quipped, 'I'm penguin dad.' (QQ's mascot is a penguin.) The other person joked back, 'Then I'm penguin mom.' That person turned out to be Wang. Pony was amused and told Wang he was a computer engineer, and this led to the beginnings of a virtual romance. The two chatted on the web, then gradually decided to take their relationship offline to meet up for real. They held a simple wedding at the Venice Hotel in Shenzhen. Pony's colleagues rejoiced because when he was single, he often worked beyond 10 p.m. Now they finally could take a breather with him starting a new chapter. Little did they know, it wasn't long before they would get bombarded again by his past-midnight emails.

In July that year, Pony's team gave him a telescope when the company moved its 760 employees to a new office, hoping their dogged leader could see the future better. They needed it, for the team was once again facing a challenging period.

Capital-market wise, Tencent suffered a blow immediately after it went public. The mobile text-messaging space was growing unwieldy. To lure users, many service providers were offering content that was borderline pornographic. Others were charging people without consent or bundling products to make more money. On the day of Tencent's listing, regulators issued rules demanding a clean-up. That tanked the stock, prompting retail investors to sell their shares on the first day, driving the company below issuance price.

Pony foresaw the dangers. During internal meetings, he told his team to colour-code content based on sensitivity. Porn, for example, fell strictly into the red camp, and he forbade Tencent to distribute such content via the QQ-based text service. Competitors were less hesitant to distribute messages that risked drawing the wrath of regulators. Tencent's revenue from text services fell to No. 4 from the top, after it took more prudent measures.

Rivals also relentlessly copied Tencent's every step. Some replicated QQ's every function, including group chat, emojis and screen grabs. They cut prices fiercely to lure more users.

More trouble ensued when telecom operators like China Mobile started asking for a larger cut. For these state-owned enterprises, the instant-messaging-based text system was creating a lot of work but not enough returns. Those services accounted for 70 per cent of complaints, yet only 3 per cent of revenue. The state-owned entities were now asking for a bigger split. Tencent had to drastically give away its share of income, going from earning 85 per cent of the cut to half.

And Tencent was about to meet its strongest competitor yet. In 2004, Microsoft-backed MSN ventured into the market, threatening to uproot Tencent with its technology and war chest.

CHAPTER 3

DAVID TO GOLIATH: A SOUL–SEARCHING JOURNEY

MICROSOFT BATTLE

In 2004, Jeff Xiong Minghua, a nine-year veteran at Microsoft, was dispatched from headquarters to Shanghai. The news triggered a sense of unease for Pony and his team. For years, Microsoft had flirted with the idea of expanding its services in China. To the first generation of tech companies, the idea of locking horns with the company founded by Bill Gates seemed an insurmountable challenge. Pony was about to meet his Goliath.

Microsoft was a formidable competitor, partly because its instant-messaging product MSN already enjoyed a significant user base in China, even without a local operation. Its user numbers were three times that of NetEase, Tencent's smaller competitor.

MSN had another edge. Its chic and streamlined interface appealed to the urban population of office workers and students, who considered QQ too grassroots and unruly for their taste. With Chinese families in cities gradually picking up in wealth, personal computers became more common in households. I was introduced to MSN by my classmates, and before I knew it, the software became the first thing that popped up on my desktop. It was running in the background when I worked, when I played music, and even when I wasn't using the computer. There was a strange sense of comfort in seeing who was logged on during the wee hours, the beeping sound of someone reaching out providing a dopamine rush.

Among the 20 million business and white-collar users in China at the time, MSN garnered a 53 per cent market share, six percentage points higher than QQ's, according to Analysys International. It surprised Microsoft's senior management and convinced them of the opportunities in China.

Jeff, a bespectacled and scholarly product manager, hailed from the land-locked province of Jiangxi, famous for being the base from which Mao conducted guerrilla warfare. He studied computer science at the National University of Defense Technology, backed by the Chinese military. It was one of the few schools in the 1980s that provided courses on the subject. He came across Windows when working for a Taiwanese company, helping localise the software for China. After studying at the Chinese Academy of Sciences, he joined a US joint venture financed by the research centre in 1993. The managers dispatched him to Hartford, Connecticut, to help expand the business, focusing on software systems.

'It was a tough two years. All the English I learned in China wasn't enough to help me get by,' Jeff said, adding that he was a one-man band working in a startup looking after everything from code testing to customer support and sales.

When he felt it was time for a change, he sent out his resumé to only two companies: IBM and Microsoft. 'For programmers like us back then, these two were it,' Jeff said. Microsoft snatched him up in 1996 after his short stint at IBM. From there, he obtained a front-row seat to the company's battle with Netscape, and helped improve the Internet Explorer browser and chat messaging service MSN.

By the time he planned to cash in his company options in 2001, Jeff had already made a name for himself among China's fledgling tech circle, and was highly regarded by Tencent's founding team. He'd written two books on software development and was a part-time lecturer at a top-tier university in China. Among people who attended his talks were two of Tencent's co-founders, one of which was Tony Zhang, who reached out.

Over two bottles of red wine at a restaurant in the posh French concession in Shanghai, they exchanged thoughts on technology, coding and the future of China's tech landscape. Jeff was impressed by the Tencent co-founder's tech know-how, but dismissed the idea of joining the team. 'At the time, MSN didn't really take QQ seriously. Their UI really sucked,' Jeff recalled.

The industry joke at the time was that QQ had done all the hard work – educating users and cultivating their habits – and MSN was coming into the market at just the right time to reap customers ready for an upgrade. QQ was viewed as a chat service for younger and lower-income people who used it to communicate while playing desktop games and seeking hook-ups in the virtual world; MSN would become the tool of choice for the white-collar community and a means for sharing information and documents for business operations.

Jeff officially relocated back to Shanghai in 2003 – bringing along his wife, elder boy and younger daughter who'd grown up in Seattle – to help Microsoft expand in China. It didn't take long for him to assemble a thirty-strong team. With his influence, he corralled talent from top-tier universities across the country. To Tencent's founding team, it was as if an aircraft carrier had moved into their waters, and they were fighting with a battleship.

In order to beef up the content provided via MSN, the company outsourced its portal business to local partners, carving out different verticals such as e-commerce, automobiles and news to companies including Alibaba. It was an astute move for Microsoft, helping it skirt the risks brought on by local regulations in content, but also monetising its messaging traffic instantly. Overnight, the US company had assembled a team of partners, forming a potent alliance against QQ.

MSN also took a page from QQ's own playbook. It acquired a local Chinese company so it could convert messages people received on their desktops to mobile text messages for a mere price of $1.20 per month. At the same time, it linked its services with Yahoo globally. Those measures posed a strong threat to Tencent.

In response, Pony's team carried out its biggest overhaul of QQ since going public. In September 2004, QQ bolstered its capacity for file sharing and storage to increase its popularity among the white-collar population. It was the first step in a series of turnarounds for Tencent. Pony proved his company could win over users with more spending power.

Pony also decided to emerge from his usual silence, laying bare to the world his vision for QQ in a media conference. By June 2005, QQ had 440 million users, the population of the US and Japan combined. He proposed that the definition of instant messaging be redefined, adding that products like QQ were no longer just tools of communication but a platform for information, entertainment, games, blogs and videos. Pony declared that chatting platforms were changing people's lives and that 'China was leading the world in instant messaging.'

Unaccustomed to making speeches in public, Pony was nervous during his debut and read from written notes. He would even blush in front of just a hundred people, colleagues recalled. Yet his predictions about how instant messaging would incorporate elements of entertainment and social media, and how users would demand better security and privacy protection, paved the way and strategy for the company over the next decade.

Pony was stubborn about one issue. He guarded QQ's moat and walled garden jealously, opposing any suggestions to open up the platform on the grounds that it was the best way to serve user interests. He turned down MSN's invitation to connect the two platforms.

He was right to hold out. Internal management issues at Microsoft in China were causing the local teams to struggle after their preliminary triumph. Due to bureaucratic structures that required information get passed up the chain back to headquarters, decision-making at MSN China became significantly slower than its competitors, who were racing against the clock.

For instance, Chinese engineers proposed that MSN users be allowed to receive messages sent to them while they were offline, the moment they logged on. That suggestion, however, didn't even qualify for discussion at headquarters. At Tencent, they managed to incorporate the feature within weeks.

'Yes, Tencent was bad with UI at the time and their service had a lot of bugs, but they were fast, and being able to provide something is better than not having anything,' said Jeff, who himself was restricted to only approving projects that required $500,000 or less, having little control over the overall business strategy.

Reacting fast, making small innovations and upgrades, is one of Tencent's hallmarks even today. At any unit, one can expect the team to provide fixes or upgrades to its services at least once every two weeks. The concept became known as 'xiao bu kuai pao, kuai su die dai', meaning taking small steps to run fast for quick evolution – a practice now widely adopted by the Chinese tech industry.

Microsoft's tradition of making software meant the company was prone to developing upgrades in a more prudent, albeit much slower fashion. Tencent being small, nimble and an internet operation from Day 1, could afford to update its services on a weekly basis and fix its bugs along the way.

Its local partners in the portal business also lacked a coherent strategy to monetise MSN's traffic. To win more advertisers, they often undercut each other, compressing ad prices. When China's regulators began to crack down on the mobile text message space, it also curbed MSN's revenue growth. Meanwhile, Tencent was sparing no effort, throwing all its weight behind QQ, which meant it could afford to look past a short-term revenue slowdown.

Another challenge for MSN was that it kept all the key user data back on servers in the US, making the service slower. The reason Microsoft made such arrangements was because China banned foreign companies from operating their own data centres, and the company refused to put its data on those provided by local providers – a sensitive issue for the government.

That meant it took as long as two minutes for users in China to log in to MSN, significantly hurting user experience. By the time it finally decided to set up a data centre in Hong Kong, which was closer to main-land China, MSN had already lost significant market share to QQ.

MSN's fatal mistake came around the end of 2005, when it incorpor-ated MSN's chat service into a larger platform known as MSN Live. To many users, the upgrade looked like a web page instead, and it appeared that the company's chat service had all but vanished overnight. When Pony's team saw the new version of MSN Live, they knew the battle was over and they had won.

And so it was that the software behemoth co-founded by the legendary Bill Gates joined a growing list of foreign tech titans felled by Chinese rivals – among them such household names as eBay and Amazon (both thwarted by Tencent's arch-rival Alibaba) and Uber (defeated and ultimately bought out by Didi). That victory gave Pony's company immeasurable street cred domestically, while establishing its bona fides with those in the know in Silicon Valley. It also ultimately led to an invaluable staff hire.

In hindsight, Microsoft never treated the war for instant messaging in China as a top priority. It was busy focusing on wrestling Google in the US. China only accounted for 2 per cent of its global business. For Ten-cent, though, Pony and his team were all-in.

'I think Microsoft ultimately lost to itself,' Jeff told me in retro-spection. Defeated, he left Microsoft, but it wouldn't be long before he emerged again.

Jeff might have been Pony's competitor, but the latter held no grudges. In fact, he was impressed by the Microsoft veteran through those years of fencing.

The two had first met in 2005, when Jeff went to Shenzhen to help acquire a company for Microsoft. Pony had reached out and the five co-founders at Tencent invited him for dinner. They talked about tech-nology, products. Jeff felt that Tencent's team was still stuck in the Stone Age when it came to concepts of tech team management – including the fact that they had no systemic approach to UI, or bug fixes.

'It was like Microsoft was in the sky and Tencent was still on earth, there was a lot of catching up to do,' Jeff said. 'I really wanted to help them, and teach them the more advanced approach used by Microsoft.'

So when Pony implored him again to join upon his departure from Microsoft, he took the offer seriously. Jeff's original plan was that he'd stay at Tencent for two years tops and help introduce some professionalism to the scrappy team, so they would at least understand best practices for how to structure a corporation, conduct user interface research and fix bugs. He even agreed to take a pay cut of about 50 per cent.

● ■ ● ■ ●

TURNING INTO GOLIATH

Jeff ended up at Tencent for eight years, watching the company grow from 2,000 people to a 30,000-strong army. It started with him shaking up hiring practices.

The first thing he did was make it a rule that he interview each and every candidate in the units he supervised, often scheduling those meetings after 6 p.m., after a day's hustle. Despite conducting the final round, Jeff would still cut a third of candidates that HR and his subordinates recommended. Word quickly spread that Jeff was a tough interviewer, and it piqued the interest of Tencent's other founders.

They were surprised by how Jeff seldom asked questions that had a definitive answer. Instead he implored candidates to think outside the box: Tencent, at the time, was headquartered in the FIYTA building in the north-western district of Shenzhen, notorious for its slow elevators. He asked interviewees how they would cut down waiting time by half, looking at how people approached and analysed the issue. Would they come up with one answer? Three answers? Would they ask about the budget, product design, or work shifts? Some would propose banning lower-floor staff from taking the elevator, others suggested doing surveys first to figure out the heart of the problem. But many just sat there, faces red, unable to utter a single word. China's education system emphasised memory and recital skills, meaning many were not cut out to be a product manager, he said. 'It's important that the people we hire have the ability to learn, problem-solve and come up with solutions on their own.'

The stringent hiring process was based on an important lesson: five people who are all-in can create better products than fifty people making a half-hearted attempt. 'You have to be extremely careful about every new hire, because that person not only affects his or her own output, but can change morale for the entire team. If they slack off and get away with it, others will look to it as a benchmark and it will drag down everyone,' he said.

The other thing Tencent was missing was a performance review system to support the rapid pace of expansion. The scrappy firm at the time had no mechanism to backstop and prevent bugs, nor any way of tracking how many bugs each of its programmers introduced, or to penalise and reward staff accordingly.

It was important to create dual career tracks for people who wanted to become managers and others just keen to be experts in their field. Tencent divided the engineers into four categories: associates, engineers, seniors and experts. Those who wanted to apply to become a senior needed to have come up with products that generated more than 10 million users; for experts, they would need 100 million.

In conjunction with these changes, Tencent also began reorganising the company to give people more freedom in pursuing new projects. It ditched the notion that only the research and development centre could generate ideas. Innovation became completely decentralised. That in turn fostered the notion that people in the company had to create a product to prove their worth.

On the upside, Tencent's staff grew militant when it came to tracking rivals and new products in the market. They were avid testers and had no qualms about ripping off concepts from competitors.

That strategy worked well in helping Tencent keep on top of trends and vanquish smaller, budding companies. With QQ's user numbers, the company leveraged the traffic and the network to outcompete them. It catapulted Tencent's performance, boosting share prices by forty times since listing to a record high by the end of 2009. Pony was feeling pretty good, so good that he bought an 8,000-square-foot mansion for HK$480 million in the scenic Shek O district, becoming neighbours with the son of Hong Kong's richest man, Li Ka-shing.

The downside of the strategy, though, was that the quality of work from teams varied and people started rushing out products. Also, it wasn't long before Tencent gained notoriety for being a killer copycat.

On a July 2010 magazine cover, *China Computerworld* superimposed the words for simulating sex acts with a dog next to Tencent's iconic penguin. That bloody image of its mascot stabbed by three knives still haunts Martin Lau.

'It created a lot of hard feelings throughout the entire industry,' the Tencent president said in an interview in 2017. 'People say whenever I have a product, Tencent is going to come into my territory and eat my lunch. This is my enemy.'

3Q BATTLE

Among all the scuffles, none put Tencent's reputation to the test as much as its battle with Qihoo, an internet security startup founded by the combative entrepreneur Zhou Hongyi.

Compact and with a receding hairline, Zhou was a contrarian who branded himself as a champion for the 'little guy': fighting for users and taking on monopolies such as Tencent.

Qihoo gained more than 100 million users via its mobile browser and antivirus software within four years of its formation. In 2010, noticing an encroaching Tencent, Zhou launched a blitz against the social media giant by accusing it of spying on users.

Qihoo claimed it had obtained evidence of how Tencent's QQ instant-messaging service monitored users and accessed data files it had no rights to.

Under the campaign, Tencent was portrayed as a lurking big brother. Media-savvy and in-your-face, Zhou at one point used his personal Weibo account, China's Twitter equivalent, to broadcast to his millions of followers his scepticism about his arch-rival.

By the time Tencent's staff resumed work after a national holiday, it was in the throes of a public crisis. The company chose not to respond and filed a lawsuit against Qihoo for unfair competition. That backfired and only reinforced its image as an arrogant behemoth. A catchy song, 'Try not to be too Ma Huateng', even started circulating online, lambasting Pony for being a copycat.

Then Qihoo delivered a bombshell on Pony's birthday, 29 October, in 2010. It offered a product called QQ Bodyguard, promising to help users identify all the security loopholes in Tencent's signature service. That night, Tencent's top executives gathered on the thirty-seventh floor of its headquarters in Shenzhen. Pony's face was ashen after his chief technology officer delivered the rundown – Qihoo was on the verge of poaching 20 million QQ users. If this lasted for just one more week, and each user had forty friends, it might lose all of its 800 million users to Qihoo.

On 3 November, Tencent informed users that QQ would make its system incompatible with Qihoo's – it was forcing users to make a choice. What went down can only be described as farcical, as Tencent's public-relations manager broke down in tears during a press conference, lamenting how reluctant the company was in forcing users to choose.

Local newspapers didn't take to the incident kindly. 'Tencent Empire Out of Control', one headline read; 'An Entity Without Respect for the Public Can't Become a Great Company', wrote another.

Tencent invited seventy-two industry experts to a series of ten closed-door meetings, which employees referred to as the 'conference of the gods'. The invitees were asked to address executives with their criticisms and recriminations – and the visitors didn't hold back.

'Tencent staff were quite clueless as to how strong and blunt the feedback would be,' said Hu Yanping, founder of the Beijing-based internet consultant DCCI & FutureLab, who took part in the sessions. 'It was the first time high-level management at Tencent faced such a fusillade.'

The feedback helped the executives understand that their problems ran much deeper than lousy public relations. In particular, the company needed something to help shake its reputation as a ferocious copycat.

Tencent's lawsuit against Qihoo over alleged unfair competition lasted four years, ending with its victory. The latter was ordered to pay 5 million yuan in compensation. The battle with Qihoo, though, served as a wake-up call for Pony.

'It wasn't the fight that mattered. Everyone was trying to leverage that opportunity to criticise us. Then we started to realise, okay, there's something wrong,' said Martin, who had by now taken on a much more important role within the company.

He was in a position to make changes. The former Goldman Sachs banker had advanced rapidly towards Tencent's nexus of power, becoming president and overseeing day-to-day operations within a year. His rising status also coincided with the exit of a co-founder. The charismatic and outspoken Jason Zeng, their original sales maven, left Tencent in 2007 to focus on investing in startups. While no whiff of a rift ever emerged in the public domain and things seemed cordial, years later I was told that part of it came down to a divergence in vision and also a crucial clash of personalities.

Pony and Jason had unspecified differences they couldn't resolve, according to people who knew both. Ultimately, Pony was the one who called the shots, due to his larger stake in the company. Martin, on the other hand, posed no threat as a professional executive and was perfect to play the supporting role Pony needed.

In an internal exercise to ensure leadership succession, each of the founders picked a person they saw as fit to take on their job. Pony picked Martin.

It wasn't just politics. The battles with Qihoo and the stress of managing a 30,000-strong workforce had also got to Jeff. By then, he was the company's oldest staffer at the age of forty-eight, compared with a workforce that averaged twenty-five. With his kids planning to go back to the US for school, Jeff decided it was time for him to move on to the next chapter of his life.

Tencent, in its journey from underdog to powerhouse, encountered the same growing pains that all companies face as well as a more pressing question: whether Pony and his creation was a greater good for business, society and the future.

Tencent had already become a heavyweight in China's tech industry, yet most of its senior executives had yet to catch up and shift their mentality. Their strategy of working on everything and trying everything to see what would stick had to change.

In stealth mode, Pony and Martin were working on a plan to demonstrate the company's resolve to change its strategy.

LANDMARK DEAL

Around mid-2013, I got wind from my contacts that Tencent was about to embark on one of its biggest structural changes to date. My source left me with the tantalising tip that Tencent was about to broker a landmark merger and acquisition deal. Over the next few weeks, I followed the clues and a series of leads.

It turned out that the acquisition involved one of China's largest search engines. Tencent was willing to pay $448 million for 36.5 per cent of China's third-largest search engine, Sogou, merging its own search product with the smaller company's in September 2013. The move accomplished two things in one stroke: it helped Tencent stake a champion in the country's competitive search industry, outbidding its fiercest competitor Qihoo, which was also in talks to acquire Sogou. More importantly, it was the first time the company was willing to forgo 'doing everything itself' and back a champion instead.

Martin said years later that the thinking behind it was quite straightforward. Tencent had two core competencies: capital and traffic volume. 'There's a lot of great entrepreneurs outside, and if you turn every single one of them into your enemy, it's not a good thing. As a company you can't actually hire these entrepreneurs, so what are you going to do? Investment is actually the best way to get a piece of their action.'

This became the breakthrough in Tencent's new strategy: from then on, the company would embark on a new chapter by becoming an incubator for startups, instead of a killer. It would invest in them, often by taking minority stakes, and grow an ecosystem of companies.

Even though Tencent never made its plans public at the time, from the outset I had started to feel the shifting tides. It wasn't just Tencent's own mentality that was changing, its presence in the entire Chinese tech industry was starting to evolve. Investors caught on and the company became an increasingly important component for mutual funds, sovereign wealth funds and retail investors who were betting their nest eggs on the idea that Tencent was only going to get bigger. At Bloomberg, my editors and I adjusted our coverage priorities and started tracking down every move that Tencent made, most of which included what it was betting on and where it was putting its money.

Five months after the Sogou deal, my colleagues and I at Bloomberg broke another important piece of news: that Tencent would sever its e-commerce unit, sell it to China's second-largest online retailer JD.com, and invest $214 million in the entity in exchange for a 15 per cent stake of the company.

Tencent figured that, since it had no clue about how to run an e-commerce business, it would simply pick one of the best players in the market and stake its interest by investing. Instead of trying to compete with every newcomer on the street, it would leverage its gargantuan web traffic and capital to support these startups.

'Tencent's investment into JD was a strategy shift, a breakthrough moment,' said Zhang Lei, Hillhouse Capital Chairman and Tencent investor. 'Before, it wanted to do everything on its own. After that, it only focused on what it did best and entrusted other sectors to partners.'

It took a significant amount of persuasion and goodwill. Smaller companies like Sogou mistrusted Tencent at the time and the company's motives. Handing its own search team over to Sogou during the merger helped ease those concerns. Internally, the decision also sent a strong message – Tencent's own search business had nearly a thousand people at the time, and its e-commerce business had close to eight thousand staff. 'When that many people actually move, people start to believe in the resolution,' said Martin. 'Internally people start to realise if I don't hold myself up to a higher standard,' it means anyone could be in danger.

Those two investments would become the road map for some 800 investments to come in the following years. Now one of China's biggest investment powerhouses on par with global funds including KKR and Sequoia, Tencent's deal-making prowess also is aided by the fact that it has a powerful network effect that startups can tap. Of those investments, at least a hundred and twenty-two have become unicorns, and at least sixty-three have gone public. The combined market value of companies in which it holds a stake of more than 5 per cent was worth more than $500 billion at one point. Those investments include companies like Spotify and Snap. It also included a 5 per cent stake in Tesla, which came after Martin called on Elon Musk in the autumn of 2014 for a tour of Space X, before another meeting in 2016.

Pony had come a long way from being an introverted computer programmer. By 2014, Tencent's market value had exceeded $150 billion, becoming the largest Chinese internet company, surpassing the first generation of local giants, including search engine Baidu and news portals NetEase and Sohu.

It was no longer the dingy little startup operating out of Hua Qiang Bei on a shoestring budget. The strategy of opening up its platform – sharing traffic and tech know-how with companies it invested in – would pave the foundation for the company's next growth spurt and a journey towards perhaps someday joining the trillion-dollar club and surpassing Facebook in size.

PART II

TENCENT UNBOUND

CHAPTER 4

SHIFTING WINDS

In many ways, the story of the rise of Tencent is that of the country's as a whole. Tencent – and an army of tech corporations that it backed – was born out of unprecedented economic expansion, a nation's growing confidence and gradual acceptance into the international community, and a billion people expecting a better life than their parents'.

I was privileged to not just witness, but also chronicle, the industry's golden years. When I gave up a job covering finance to join Bloomberg in 2012 as a technology reporter, colleagues and friends couldn't fully comprehend my decision, thinking it a step down as the hottest stories about China were centred on banking back then. The eyes of the world were trained on what they regarded as a global disaster-in-the-making: a runaway shadow banking system and local government vehicles that funded white elephant projects.

Back then, China's tech industry had yet to come into its own. It was two years before Alibaba would go public – a coming-out party not just for the e-commerce giant but, in many ways, for the entire domestic tech sphere. Silicon Valley was the undisputed centre of the tech universe, where names from Facebook to Amazon commanded global adulation and the lions' share of international investment capital.

China was a tech follower, populated by copycats such as Weibo (Twitter), Baidu (Google) and Alibaba (eBay). Tencent's WeChat was in its infancy, and ByteDance's TikTok hadn't even been conceived. There was still the notion that for tech and computer science majors, the moment you graduated, you were in danger of going jobless. The iPhone had emerged only a few years previously and the desktop still ruled, at least in China. The country's tech players were in awe of their American peers, aspiring towards and looking up to the likes of Yahoo, Microsoft, Cisco.

How times have changed. Nowhere was that transformation more evident than in the capital of Beijing itself. That national psyche is reflected in the speed of change the capital constantly undergoes. Every few months, it's stippled with a new skyscraper, subway station, night club or luxury residential complex with names like Palm Springs or Park Avenue. In a city where the pollution can be so thick with particles that the act of breathing is akin to eating air, the willow-flanked streets spewing catkins in spring are bulldozed every few months and reborn with monikers like Innovation Street.

The landlocked city still shrouds you in the scent of sulphur dioxide and coal the moment you step off the plane. For those passing through, it presages coughs and rhinitis; some would joke 'ten years off your health.' But to me it's always the scent of home – burning stubble in the wheat fields outside the Fourth Ring Road, where the Olympic Bird's Nest Stadium now stands; kebab stalls tucked in the hutongs around the iconic Drum Tower, fanned by grizzled migrants with pastel-hued blow-dryers; whiffs of petrichor washing away the humidity after the first autumn rain; and tobacco-choked night clubs near the Worker's Stadium, where Lamborghinis moonlight as Uber rides to pick up girls. One of my foreign friends calls it 'the developing economy scent'. To me, in an ironic way, it's the scent of hope and ambition.

Over a heady decade, I've had a front-row seat to a tidal wave of disruption and evolution that I could not have foreseen, spearheaded by some of the brightest and most tenacious minds of our times. There were the in-depth sit-downs with Jack Ma in New York right before Alibaba's record 2014 initial public offering; tracking down bitcoin miners who resided in co-living communes back before WeWork popularised the concept and the cryptocurrency was at $200; unearthing the secret formula of the nation's top e-sports champion (he trained for sixteen hours a day before retiring at twenty-one years of age); interviewing hackers who were creating tools to track censorship on social media and helping Chinese people circumvent the Great Firewall. Tech became one of the best lenses through which to understand the aspirations, ingenuity and tensions behind the world's second-largest economy.

There was never a dull moment. The narrative was constantly changing, underpinned by the morphing technology, personalities and policies that collectively shaped how a billion people perceived the world, conducted their relationships and lives, and forged their own identities. It was then that I first grasped the power behind Tencent's WeChat. A tweak in its algorithm could drastically alter the lives of hundreds of thousands of content creators who lived and died with the attention economy, funnelling traffic to or away from their works.

That was just one example – powerful internal and external forces were constantly reshaping the economy and industry. A clandestine meeting could lead to multi-billion-dollar mergers and acquisitions, sometimes within the span of days, affecting the lives of tens of thousands of workers and avid product users. A whiff of policy tightening from Beijing could tank the price of the world's largest companies and cause hundreds of billions of market value to vanish in the blink of an eye.

In China, three companies developed the scale and resolve to weather those changes. Internet search giant Baidu, Alibaba and Tencent, collectively known as 'BAT,' became the first-tier triumvirate that anchored an entire industry. Of the three founders, Baidu's Robin Li came with the most accomplished pedigree. The only US-educated entrepreneur among the three, Li had racked up an impressive list of academic achievements, earning a bachelor's degree in information management and a Master's at the State University of New York at Buffalo in 1994. Robin was a 'Hai Gui', sounding like the Chinese phrase for tortoise and referring to people who received an overseas education and returned to the motherland, a class of people who commanded undue respect in the days before China started exporting 700,000 students a year.

Pony and Alibaba's founder Jack Ma came with far less shine by comparison, and were jokingly referred to as 'tu bie' or local softshell turtles, a phrase not unlike the Western 'bumpkins.' And yet, it would be these two 'tu bies' who would dominate as the country transitioned into the mobile era. They managed to outshine Robin, surpassing him in the span of less than a year, underscoring the cutthroat competition and the importance of acting fast. Witnessing giants like Baidu and Yahoo become pale shadows of their former selves is also what keeps the two Mas on their edge.

The duopoly they created defines China Tech Inc's past decade. With the backing of the two companies and billions from global venture and private equity funds, a new army of younger companies flourished. The amount of money Chinese-focused venture and private equity funds raised exploded nearly four-fold to $120 billion at its peak in 2017, according to London-based researcher Preqin. That bounty helped China transform from industry backwater to one of the most dynamic and coveted markets on the planet. Once derided as copycats, companies like Alibaba and Tencent became known as innovators in mobile payments and messaging.

Both joined the rarefied ranks of the world's largest companies and became mainstays in investment portfolios across the globe. Their founders achieved local cult status and international renown – Jack Ma, in particular, and his folksy wisdom became fixtures on the global conference-speaking circuit.

In contrast, Pony shied away from the limelight, yet his company became synonymous with Chinese innovation over the years.

The greatest surprise was Tencent's innovation with WeChat. And even more so, the patience Pony showed in monetising the product. One of the first assignments I got covering Tencent was an article inspecting WeChat's business model. Analysts and investors at the time were skeptical that the product could figure out a viable path.

In the process of that, I was keen to figure out just what functions avid customers were using. At the time, one of the most popular was the 'drift bottle' service that helped people find friends. WeChat let users throw or pick up a random audio or text message. The interface has a picture of a stretch of beach under a blue sky, where users can pick up a bottle as a virtual hot-air balloon drifts overhead. The user can respond to the message or throw it back. I spent one night frantically throwing virtual bottles trying to see if anyone would tell me the appeal of connecting with strangers, and if those dialogues lead to longer-lasting relationships in the real world. Turns out it became one of the first versions of a hook-up app, predating Tinder by two years.

In the years to follow, WeChat would introduce functions that pushed it beyond a simple messaging tool to become the one app that rules them all – 'Moments', a social media feed that allowed users to share information, photos and videos; 'Official Accounts' that enabled millions of content creators and companies to broadcast their ideas; 'Mini Programs' that allowed users to access tens of thousands of other lite apps on WeChat's interface without having to download the full versions; and now 'Video Accounts', its means of countering TikTok in short clips.

Just one product has ignited a plethora of companies and encouraged aspiring young people to create their own ventures, whether it be leveraging WeChat's traffic or trying to topple the giant.

In the beginning, Tencent focused on mobile games while Alibaba strove to dominate commerce. Pony's company at one point made more than 70 per cent of its revenue from games. As of October 2021, it has backed more than a hundred and eighty gaming companies globally since 2008, according to Niko Partners, the largest of which included Riot (the brains behind League of Legends), Activision Blizzard (Call of Duty, World of Warcraft), Epic (Fortnite) and Supercell (Clash of Clans). That impressive portfolio makes Tencent the largest gaming factory in the world. It also helped the company leapfrog from being a simple Chinese distributor to becoming a respected publisher and creator of top-notch mobile titles. That would prove especially important as gamers globally migrated from desktop PCs and consoles to smaller phone screens.

As they shored up their dominance of their respective spheres, both would gradually become industry kingmakers. Tencent's and Alibaba's decision to set up venture investing arms to back the next generation of startups was one of the key drivers of their eventual success. Their motive was two-fold. They could give money to companies purely based on projections of potential financial return. But more importantly, they were both looking for businesses that would strategically benefit and propel their rapidly growing ecosystems.

While both companies say that they are enablers and respect founders, the style and tactics they deployed during these investments were drastically different. When I talked to startup founders, they often described Alibaba as much more aggressive when it came to giving direction and maintaining control to ensure that their strategy aligned with Alibaba's objectives. Tencent, in comparison, was much more hands-off in its approach, but also very guarded when it came to sharing resources, especially using traffic from WeChat to help promote other services. It wasn't because WeChat was stingy. Even within Tencent, internal departments needed to fight to prove their worthiness before the instant-messaging app was willing to work with them. Pony gave WeChat's founder Allen Zhang a great amount of freedom in letting him run his product and business group as he saw fit, meaning WeChat was a little fiefdom unto itself.

Among the hundreds of startups that Alibaba and Tencent have invested in, there were two that would prove particularly crucial in helping to connect hundreds of millions of users to real-world services in the mobile era. They were food-delivery giant Meituan and ride-hailing firm Didi. And, as their rivalry played out, both would become a key battleground in the grand struggle between Alibaba and Tencent.

Corralling both startups helped lay the foundation for Tencent's success in the age of the mobile internet. The story of how Tencent made those decisions, outwitting its largest rivals and coping with regulators for a decade or more, is what defined the company that at one point became larger even than Facebook.

CHAPTER 5

ALIBABA THE ARCH-NEMESIS

It's hard to talk about Tencent without bringing up Alibaba. Together, the twin pillars of China's internet industry dominate the landscape.

Asia's largest companies have been jostling with each other for well over a decade, a contest that accelerated after Alibaba pulled off the world's largest initial public offering in 2014. When Tencent's founders say the competition makes them stronger, Alibaba is the one that pushes them the most.

In their struggle to maintain the upper hand, the two companies led by Pony Ma and Jack Ma have both expanded beyond their core business, stretching the frontiers of their conglomerates to everything from online finance to entertainment and cloud computing.

Alibaba's story is also important because without it, there wouldn't exist the Chinese online payments industry we see today – by most accounts the world's broadest and most sophisticated digital financial system. Its e-commerce battle with eBay spurred the growth of Alipay, which in turn set the stage for a fearsome rivalry with Tencent in the mobile payments era.

Today, Alibaba's might is on full display in the eastern metropolis of Hangzhou, a city of 10 million people it calls home, a favourite of foreign tourists for its picturesque West Lake, ancient stone bridges and historic sites. That power radiates out from its sprawling campus near the wetlands of the Xixi national park to envelope just about every large city in China.

From the cab one hails to the co-sharing bike you rent, from the mobile payment QR codes on display in millions of convenience stores and malls to online movie tickets and blue-coated food-delivery men shuttling about on scooters – a plethora of services today are backed either by Alibaba's money or its cloud infrastructure.

Nowhere is that prowess more obvious than on 11 November every year, when the company presides over Singles Day, the world's largest shopping festival, selling more merchandise than Black Friday and Cyber Monday combined.

In 2021, an estimated 900 million-plus shoppers from China to Russia and Argentina scooped up $85 billion's worth of iPhones, Dyson appliances and Ugandan mangoes over a two-week period leading up to the 11th, more than the GDP of Estonia or Nepal. At its peak, Alibaba's servers process more than half a million transactions a second across its sites, delivering over a billion packages in a single day. Hundreds of journalists and investors typically swarm the company auditorium, watching satellite and 3D images of deliveries whistling across the globe.

It's a mammoth undertaking few other companies can pull off. Alibaba's staff pull all-nighters, often crashing on office floors and holing up at their workstations for a week without going home. Staff, mostly in their mid-twenties, ride free-to-rent bikes between the gleaming buildings in white and orange, Alibaba's signature colour.

In 2019, Taylor Swift and Chinese piano maestro Lang Lang headlined a concurrent gala at the Mercedes-Benz Arena, about a hundred miles north-east of Alibaba's command centre. The show, co-produced by entertainment companies backed by Alibaba, drew 100 million viewers over four hours. Over the years, the same show has featured Nicole Kidman, Daniel Craig, Kobe Bryant and Scarlett Johansson.

Local influencers go on a seven-hour marathon hawking products ranging from eye-makeup and facial creams, to jeans and electronics. Between the top two live streamers, they managed to sell 7 billion yuan of merchandise in 2019. Grandmas, office workers and students watch the livestreams, then scoop up Bordeaux wine, UGG boots, SUVs and high-end Japanese toilets from their beds, couches, dinner tables or during their commutes to work.

'If Steve Jobs created the operating system for the smartphone, Jack Ma and his team created the operating system for commerce in China and the future,' said Porter Erisman, who directed the documentary, *Crocodile in the Yangtze: A Westerner Inside China's Alibaba.com*, chronicling his eight-year employment at the company starting in 2000. 'He's someone who loves a challenge. He's motivated by doing things that push China forward.'

The only company that can rival Alibaba in influence is Tencent. Yet the two companies and their founders' personalities couldn't be further apart.

Alibaba's Jack Ma made a name for himself globally for being the brash, outspoken and impossible-to-ignore entrepreneur who hobnobs with celebrities at Davos and doles out pithy pearls of wisdom such as 'We are never short of money. We lack people with dreams, who can die for those dreams.'

When he shows up at conferences, young people rush to the front of the auditorium, calling him 'teacher Ma' in affection, something I seldom see at forums attended by Pony. Jack can be eccentric as well. Embracing his celebrity status, he's dressed up as Michael Jackson and a blonde girl in a Princess Peach-like dress. He once turned up as a faux rock star in red leather tights, shades and a mohawk at Alibaba's own staff celebrations.

Yet that outward exuberance, a constant craving of the spotlight, eventually landed the billionaire in hot water in October 2020, when the Alibaba co-founder infamously railed against the inefficiencies of China's traditional finance industry and its regulatory overseers during a prominent business forum in Shanghai. It was a stunning moment, a self-made Chinese entrepreneur openly defying his political masters and criticising the state-run enterprises that form the bedrock of the world's second-largest economy.

The fallout was swift and unprecedented. In November that year, the country's stock market watchdog pulled the plug on the $35 billion initial public offering of Alipay parent Ant Group Co., the company Jack created by spinning the payments service out of Alibaba. In the months that followed, Ma vanished from public view as regulators pursued a relentless campaign against both Ant and Alibaba, launching an antitrust investigation against the latter e-commerce giant that ended only with a record $2.8 billion fine. In the months that followed, Beijing's autocrats fired regulatory salvo after salvo at the two firms, issuing decrees on a weekly basis that seemed intended to curb their once frenetic pace of growth – from demanding Ant streamline and restructure its sprawling structure to ordering Alibaba to halt its years-long practice of securing merchant exclusivity. The assault eventually spiralled beyond Jack's companies to encompass the entire internet industry – ensnaring even Tencent when separate agencies began to order curbs on mobile gaming. All told, Tencent, Alibaba and their largest industry peers lost more than $1.5 trillion of market value collectively at the height of the crackdown.

It's an episode Jack Ma would no doubt prefer consigned to history – a mis-step in a storied career with grave consequences for the entire Chinese technology sector. To understand how we got to that point, one needs to trace the origins of the diminutive but charismatic former schoolteacher whose chutzpah and unshakeable self-confidence had worked wonders over the years.

Jack was born on 10 September 1964 to Chinese traditional musician-storytellers in Hangzhou. As a teenager, he met foreigners just as China became a tourist destination again following the Nixon–Mao rapprochement of 1972. For nine years, the youth hung out outside the Hangzhou Hotel, now a Shangri-La, getting up at 5 a.m. to talk with travellers.

Ma failed China's national university entrance exam twice before he was admitted to what is now Hangzhou Normal University. He graduated in 1988 and spent five years teaching English at a local university, earning $15 a month.

The online world opened to Jack on his first trip to the US in 1995. While visiting a friend in Seattle who showed him the internet, Jack typed the word 'beer' into an early web search engine and couldn't find any Chinese information. He decided to fix it.

'I searched the word beer, b-e-e-r, very simple word,' Jack recalled in a television appearance shown in the documentary. 'I found American beer, German beer, and no Chinese beer. So I was curious. I searched "China", and all search engines said no China, no data.'

The first step to building an online empire began when Jack registered a website, China Pages, a yellow pages-like directory, in the US.

Jack said he asked his friend to help him create a home page in Chinese. Within hours of posting it, he received five emails from the US, Japan and Germany seeking more information.

'I was so excited,' he recalled. 'I said, "This is something interesting."'

To start China Pages, Jack took 7,000 yuan from his savings and borrowed money from his sister. His first two employees were Jack's wife Zhang Ying, his university classmate and then colleague when he taught English. Zhang played an important role in Alibaba's early days until she turned to more domestic responsibilities and raising their son.

Jack's early attempts with China Pages didn't take off, so he joined the Ministry of Commerce in Beijing, helping the agency set up a website. It was there that he met a first-time visitor to China, Jerry Yang, the co-founder of Yahoo who would – years later – invest $1 billion for 40 per cent of Alibaba. The collaboration would eventually end up mired in controversy, after what Yahoo called a unilateral spinoff of the Chinese company's PayPal-like online-payment business Alipay. But Yang's patronage was exactly the early boost Jack needed – even if it took several steps to get there.

In 1999, the internet stock boom gripped Wall Street. Jack, feeling the change as far away as China's capital, left government service and returned home to tackle a new idea. Back in his Hangzhou apartment, with his wife and a group of friends, Jack set up Alibaba.com, a site that allowed businesses to sell to each other. Believing that the moment would become of historic value, Jack had the kick-off meeting recorded on film.

'INTERNET DREAM'

'Don't worry, I think the internet dream will not die,' Jack was filmed saying at the February 1999 meeting, which was later aired by state-run broadcaster CCTV. 'The reward we will receive for the price we pay in the next three to five years is not this apartment, but fifty of these apartments.'

Alibaba.com, which had struggled to find a way to generate revenue, learned that merchants were willing to pay for better displays of their goods. The website attracted 1 million users in 2002 and became profitable that year, according to details released by the company.

That year, Tencent's Pony popped up on Jack's radar. Jack, who is deeply influenced by the works of Hong Kong novelist Louis Leung-Yung Cha (otherwise known as Jin Yong) pioneered the practice of hosting technology forums in China. He dubbed these conferences Xi Hu Lun Jian, which translates literally as Swordsmanship by the West Lake. He aspired to corral the best tech minds in the country to share and spread ideas about how technology could transform the world. The industry, however, was struggling in 2002, hence most A-list CEOs chose to focus on their business instead of attending the meeting. As a backup plan, Jack invited Pony as a guest speaker instead.

A modest and somewhat shy Pony was happy to announce that four-year-old Tencent had broken even. He then said that he hoped that QQ could become a lifestyle. Pony played his cards close to his chest – or perhaps lacked Jack's flair for grandiose visions – focused on talking about his product. Little did Jack know that he and Pony would clash in the years to come.

Back then, it seemed that Jack's and Pony's businesses would have little overlap, so Jack aimed his ammunition at a much more immediate rival: eBay.

To counter eBay, Alibaba started Taobao, a platform where individual sellers could trade with each other, in 2003. Taobao publicly declared war against eBay, and the resulting media coverage drew attention to the website.

'By generating a public-relations battle with eBay, there would be a lot of buzz about Taobao, and therefore with every dollar that eBay spent, it would help feed this buzz about Taobao,' said Erisman, the former executive. He cites Jack's professed interest in Chinese martial arts as a motivating strategy: using an opponent's strength against him.

'EBay is a shark in the ocean; we are a crocodile in the Yangtze River,' Jack was quoted as saying on numerous occasions at the time. 'If we fight in the ocean, we will lose, but if we fight in the river, we will win.'

It was Yang's investment that proved instrumental in the battle. Using search-engine technology that came with Yahoo's stake in 2005, Alibaba made Taobao listings free for merchants. EBay, which continued charging for listings, announced in December 2006 that it would close its unprofitable site in China. That singular victory marked a turning point for Alibaba and its hard-driving founder. Just as Tencent established its credentials by taking on and vanquishing Microsoft in China, so too had Alibaba demonstrated its ability to grapple with the best of the West – and win.

ALIPAY DISRUPTION

One of Jack Ma's biggest creations and weapons in the fight with eBay – later allowing him to dominate e-commerce in China – was Alipay, an online payments service that enabled trust between buyers and merchants when buying goods on the internet was still a novelty. The business was so crucial when it came to winning over mobile services and users that Tencent would later create its own equivalent.

It all started in 2003, when a lanky and bespectacled accounting graduate Ni Xingjun showed up at Jack Ma's Hangzhou apartment for a job interview. In a suit with briefcase in hand, Ni looked like an insurance salesman in a room-full of people in shorts and slippers. But Ni had taught himself computer science and later wrote a thesis on data authenticity, impressing Jack and the other founders. And there in the smoke-filled apartment smelling of sweat (they worked there round the clock), he landed a job.

Ni's first challenge was overcoming the issue of people not trusting each other online. The team studied PayPal, and they considered issuing virtual coins like Tencent did, but decided it was not a viable option given China's circumstances (where both trust and basic infrastructure was lacking). Ni and his team figured customers cared most about receiving their goods, so the best option was to create an Alibaba-managed escrow account that users paid into. Alibaba held on to the money and acted as a middleman. Only after buyers got their goods would the sellers get paid.

It was a gutsy gamble for a fledgling online mall. Alipay logged its first transaction in October 2003 when Cui Weiping, a student in Japan, listed a Fuji camera for 750 yuan on Taobao. Jiao Zhenzhong in the heartland city of Xi'an expressed interest. After haggling for hours, they decided to try out Alipay's new escrow service. Jiao paid Alibaba the money and Cui, on the other end, only received the payment after the camera was delivered – and Alipay was born. That maiden transaction was marked as '200310126550336', printed out and immortalised today on a wall of Alipay's parents' headquarters.

It was all the more remarkable because Alipay's original team consisted of one finance manager, one accountant and one teller. They used Microsoft Excel to document all the transactions, manually.

During the early days, whenever staff helped complete a transaction, a light on their desk would glow and the person would stand up to accept applause from co-workers. It was also in those days that Alipay operated in a legal grey zone. Private companies were not allowed to venture into finance. Technically, online payments could be deemed illegal by the government, so Jack and his team were walking a fine line.

To encourage his staff to forge ahead, Jack would tell them: 'If someone has to go to jail, I'll go.'[1]

Transactions started to pick up. Merchants found businesses often grew when they adopted the service. Soon 70 per cent of merchants on Taobao had signed up.

Ni himself was running on little more than adrenaline every day. He blasted out a message to Alibaba colleagues whenever transactions hit 100,000 yuan a day, then lifted the threshold to 1 million yuan, and as a joke, promised to run across the office floor topless if transactions hit 7 million yuan a day. It didn't take long. He honoured that bet in the first half of 2005.

That said, Alipay could be a hard place to work.

Exponential growth brought problems. The team was stretched thin and customer complaints surged. In January 2010, Alipay staff gathered at the People's Hall in Hangzhou after a year of hard work for their annual

---

[1] 由曦. (2019). 蚂蚁金服. 中国中信出版社.

party. When they arrived, they noticed something was amiss: no party banners, no welcome drinks or dance troupes. The venue seemed solemn and grave. As they took their seats, the speakers boomed. 'You folks are killing me', 'I will never use Alipay again.' It was an audio recording of customer complaints – a litany of them. Then Jack Ma himself stepped onstage.

'What I really couldn't tolerate was the customer service experience,' Jack said, his right hand pointing towards the sky. 'I'm not grumbling, but we need the courage to admit when we're wrong, and to pay the price for the mistake.'

Ni knew what Jack was referring to. The pursuit of growth meant staff were turning a blind eye to service quality. User numbers had hit 270 million by then, 82 per cent of the US population.

Yet Alibaba's biggest controversy – which erupted just a few months after that sombre affair – would involve not service but its biggest shareholder at the time.

In August 2010, Jack moved Alipay out of the e-commerce giant and into a separate company he controlled. Yahoo! Inc. was furious, saying it wasn't informed in advance. Jack said later that he was concerned that the government wouldn't allow foreign investors in third-party payments companies – a regulatory risk that didn't materialise – and did what he felt was necessary at the time. But his reputation was damaged as investors accused him of effectively stealing Alipay.

Regardless, the stratagem appeared to work. Alipay soon won a payments licence from the central bank, helping it step out of the grey zone and become a legitimate business. Jack struck a deal with Yahoo later to share its profits.

QUICK PAY

Meanwhile, the Alipay team worked to address its fundamental issues. To address complaints from users, Alipay once again innovated with a product called Quick Pay. It created a payments ledger by linking 200 Chinese banks to a network that would speed up transactions with

Alipay. In a sense, it began carrying out the work of a clearing house or central bank.

The Quick Pay service went on to establish the model for online payments in China, including for Tencent, which later created its own equivalent – WeChat Pay.

'If we didn't take the first step to creating Quick Pay, there would be no WeChat Pay or mobile payments, and it would be challenging for other third-party payments services,' said Ni.

But playing part of the role of a central bank planted the seeds of the regulatory scrutiny that would come back to haunt them years later. Back then, though, Jack and his team had a much more immediate problem with the advent of smartphones.

MOBILE ERA

Around 2012, Alipay's staff began feeling acute anxiety over the need to have a mobile strategy. Tencent had already come up with a formidable instant-messaging app called WeChat that became an instant hit. With the arrival of 4G and greater broadband, a flurry of local homegrown apps began emerging.

'It was painful,' said He Yongming, a senior executive who helped Alipay navigate the mobile transition. 'When we shifted from PC to mobile, we were giving up a familiar technology to explore something new, yet we still decided to go all-in.'

Alipay was again in uncharted waters. Its executives were divided over QR codes and near-field communications (NFC) as the best approach to win the market. NFC was favoured because some staff thought it was a superior technology, but they didn't take into consideration how expensive the machines involved would be, nor how much merchants and retailers would balk at adopting new technology. While Alipay was busy with internal debates, it got ambushed by Tencent in 2014.

ERA OF MOBILE PAYMENTS

In China, elders and senior executives are often expected to give out red packets that contain a token amount of money during the Lunar New Year. It was a tradition within Tencent to have Pony hand out red packets every year. As the company expanded in size, lines would form for hours outside Pony's office.

WeChat staffers, hoping to come up with a more effective solution for Pony, created a digital red packet service where they could be doled out to individuals peer-to-peer or simply via a group. The company, realizing how popular the function got, rolled it out to the public within a few weeks in January of 2014 and also gamified the practice, so some people got a large chunk of money, while others got mere cents. It became an instant hit.

The feature grew so popular that the team requested Tencent deploy ten times the number of servers originally planned. During that one holiday, more than 75 million red packets were digitally sent. Tencent, wasting no time, allowed users to start linking their bank cards with WeChat, so they could deposit even more money into their accounts. By the time Alipay realised what had happened, WeChat had already become a significant competitor in mobile payments. Alipay, which took a decade to establish its dominance in the payments space, branded the event the 'Pearl Harbor Incident', underscoring how Tencent had pulled off a deadly – potentially fatal – attack.

Alipay was so flustered that it decided to switch tactics. It went after Tencent, trying to expand its digital wallet to include social media functions. That turned into one of the biggest public relations fiascos at the company.

The idea was that Alipay would add a social media function called 'Circles' to allow users to chat and form interest groups on the app. Alipay's team thought that since WeChat was able to encroach into payments despite being a messaging app, the reverse would work as well. But it failed to recognise one crucial difference. On WeChat, most of the social connections were based on real-life connections, hence there was already some level of trust when people sent money to each other through the app. Users of Alipay, on the other hand, had no real-world links with the people they added as friends through the new service. Within hours of launch, pictures of scantily clad young women began appearing on the app, in exchange for money. 'Circles' was flooded with photos of women posing suggestively in bikinis and underwear, with some indicating they were looking for hookups.

Alipay quickly removed the photos and blocked accounts that were posting inappropriate content. But the damage was done. The incident sparked a public debate about whether Jack and his company were growing desperate to increase the popularity of the app.

Lucy Peng, the executive chairman back then who oversaw Alipay's parent, lamented the damage done to the company's brand. Alipay had till then built up an image of being a serious service that people could entrust their money with. The fiasco marked Alipay's last serious attempt to expand into social media.

Internally, Alipay revisited its strategy. Ni, who championed QR codes and came up with the idea to print them out on paper or plastic sheets for merchants, decided it was time to invest in that concept. The companycreated a massive campaign, printing out QR codes for China's largest grocery chains. The codes literally cost nothing for merchants to adopt. It was much cheaper than maintaining nearfield communication devices. Alipay managed to scale fast. The tactic worked so well that it's become the standard for mobile payments in China, with even street beggars flashing barcode-like displays when asking for money.

Mobile payments changed everything. It became one of the prime battles between Jack and Pony over the ensuing years. It introduced a range of services including food delivery and ride-hailing, helping create a new cohort of internet companies, Meituan and Didi.

Jack and Pony saw the two services as a key to increasing market share in mobile payments and expanding the influence of their empires. The story of that struggle would put Tencent's determination to expand its platform and ecosystem to the test.

CHAPTER 6

THE BIGGEST COUP: MEITUAN

In Beijing and Shanghai, it's often cheaper to have food delivered than to get it yourself. Back in 2019, you could order roast duck from a local diner for $2.99, about 80 per cent less than at the register, via delivery app Meituan. You can get a 40 per cent discount on two pizzas topped with golden potatoes and barbecued seafood. Meituan charges $1.46 for a bean curd dish from another shop, a little over a third of the price on the restaurant's menu.

Life in the capital can be hard – the blackened air, the bitter winters, the three-hour traffic congestions, and the government bans on Instagram and Snapchat. But it comes with perks: burgers, noodles, grapefruit, spicy wings and cumin meat skewers delivered at 2 a.m., usually within thirty minutes.

Across the country, millions of people order two or three meals a day, as well as groceries, office supplies, haircuts, massages, and whatever else they might fancy. This $100-billion delivery market isn't exactly driven by efficiency, though, but by a fight between Alibaba and Meituan – one of Tencent's single most important investments.

Walk past any university campus gate, residential complex lobby and officer tower entrance and you'll see a band of brightly liveried delivery people impatiently stomping their feet in winter cold, rain and smog. They can be spotted shuttling through the streets at any moment of any given day, zigzagging through city traffic, donning jackets of yellow from Meituan or blue for Alibaba.

Alibaba and its various subsidiaries dominate the country's online retail market for physical goods, but Meituan is trying to lead the way in services. Its namesake app, a sort of mashup of Grubhub, Expedia, MovieTickets.com, Groupon and Yelp, has 9.5 million delivery people serving 690 million customers a year in more than 2,800 cities. Up until about late 2020, Alibaba bet it could undercut Meituan – and vice versa – in a duel to the death. Both companies spent billions in an escalating war of subsidies that might have persuaded even Amazon's Jeff Bezos, infamous for splurging to squeeze out rivals, to cut his losses.

The story of how Meituan became the largest food-delivery company, and how its founder Wang Xing became Jack Ma's arch-foe, is full of twists and turns. The company's ascendance is one of the most important chapters in China's internet history, laying the foundation for one of the biggest conflicts over the years – the food-delivery and on-demand services war between Alibaba and Meituan – as well as one of Tencent's greatest victories over its nemesis.

At the centre of Meituan was its young founder Wang Xing, one of several up-and-coming entrepreneurs Tencent bankrolled in a proxy war with Alibaba. A programmer by trade, he needed just five or six years to dominate the market for meal delivery, containing the mighty Alibaba through a combination of aggressive salesmanship, no-holds-barred discounting and incredibly effective algorithms that deployed hundreds of thousands of delivery personnel with a precision and efficiency that surprised even Alibaba, at the time the acknowledged king of logistics in China.

What made Meituan and Tencent's victory all the sweeter was that Wang had taken money from Alibaba – but when push came to shove, the ambitious entrepreneur defected to Tencent's camp.

The golden rule in China's tech landscape was that smaller start-ups had to take investment from one of the two giants: Alibaba or Tencent. Taking money from either of the two often guaranteed support in internet traffic and resources, but also threatened to typecast your startup and polarise future investment.

'It's a little bit like the Godfather Don Corleone saying, "I'm going to make you an offer you can't refuse,"' says Andy Mok, a senior research fellow at the Center for China and Globalization. 'If you don't take their money, and they invest in a competitor, it can be deadly.'

Initially some suspected an easing of tensions behind the reason Alibaba would allow one of its underlings to take money from an arch-rival. What few realised was that Meituan's deal with Tencent would turnout to be the coup of the decade for China's internet sector.

Early on, Wang felt his biggest investor at the time – Alibaba – was trying to push him around and wanted to make it clear that he ran the show. So he met secretly with Tencent in Beijing and hammered out a separate deal with the social media giant, which then became a significant shareholder in his company.

Wang and Tencent effectively pulled the rug out from under Alibaba, sealing a fateful alliance that would set the stage for a showdown in China's internet arena.

I sat down with Wang Xing on a cold December day to discuss his journey towards creating the world's fourth-largest startup. We delved into extensive detail about how the shift in alliances helped Tencent outsmart Alibaba in the mobile payment wars. And most importantly, how he and Tencent engineered that epic coup.

This wasn't the first time I had interviewed him. But in the years since we last talked, Wang had shed the persona of a reticent programmer. At forty, he was richer than NBA Dallas Mavericks owner Mark Cuban.

Wang had a reputation for being borderline arrogant. Others would say he had no ego, that the uneasiness he imparted stemmed from his icy logic and somewhat awkward personality. He's the Chinese version of Spock. When Wang talked to you, you needed to prove to him you were on top of your game. Know your facts. Follow his train of thought. Charm seldom worked.

It took years of haggling, bombarding his communications team and exhausting sources. Wang finally agreed to meet in 2018. His company had just conducted Hong Kong's second-largest tech initial public offering of the year.

Located in the eastern financial district of Beijing, Meituan's headquarters, nestled in a bustling industrial park, boasts sparkling office buildings and expansive green lawns, through which food-delivery men scurry constantly on electric scooters. Coders and salespeople hunker over open desks decked with company mascots and asparagus fern. Many take a break from their long work hours during lunch, napping on camp beds and inflatable pillows. Chinese tech companies are notorious for their gruelling hours, often driving staff to work a so-called 996 schedule: 9 a.m. to 9 p.m., 6 days a week, including holidays. Meituan was no exception.

At Meituan, there's also one special office. The placard on the door reads 'internet security police room'. It's the infamous setup that allows Chinese public security forces to intervene in company operations whenever the government wanted to roll out a political directive, scrap a message or track certain users. Until 2015, these operations lurked as uneasy urban legends among internet circles before they were publicly disclosed. While these rooms often run empty with no one on duty, the power they evince isn't taken lightly.

I wanted to know how people like Wang, whose grandad had endured persecution during the great upheaval known as the Cultural Revolution, rose to become billionaires. Wang's life echoes his country's tumultuous transition from a poor, agrarian nation to economic powerhouse. Private enterprise was illegal up until the year before he was born. I was curious what his relationship with the Chinese government was like; how he learned to cope with censorship and suspicion from the country's ruling elite; how living behind a sanitised web moulds an entire generation; and whether what happens in China today could herald an alternate future for our world.

Wang Xing sat in a room filled with overused markers and white boards scribbled with indecipherable notes. He wore an island-white button-down, dark jeans and wire-rimmed glasses. While his demeanour had morphed over the years into something more befitting a corporate chieftain, one thing stayed the same – his buzz cut narrowed to a widow's peak accentuating his large forehead, which many Chinese people regard as a symbol of intelligence. He still comes across today as more of an awkward programmer than a freshly minted billionaire commanding a workforce of 90,000. But once he opens up, he doesn't hold back and can be fiercely critical and outspoken.

CIRCUITS AND COPPER SULPHATE

Wang Xing was born in 1979, at the outset of Deng Xiaoping's reformist regime. China had just recovered from the turmoil of a decade-long Cultural Revolution under Mao Zedong. Growing up on the poorer side of Fujian – a coastal province known for its undulating tea plantations and hard-charging businessmen – Wang got his first taste of entrepreneurship by observing his dad.

His father helped raise three younger siblings at the age of sixteen. Wang's grandfather, a Chinese opera playwright, committed suicide within the first year of the Cultural Revolution. Wang didn't learn about this dark chapter until his grandmother, an intellectual who graduated from Xiamen University, passed away in 2012. 'I guess he was sort of persecuted,' Wang said with a sombre air.

Wang's father attempted everything from mining to gardening, before finally taking off with a cement business in 1992. That enabled Wang and his elder sister to enjoy a relatively well-off childhood. Wang would tinker with circuits and copper sulphate crystals in the bathtub. He also ditched piano lessons to get an early start with computers.

As a child, Wang played the first-person shooter game Wolfenstein, trying to assassinate Hitler. He learned English and history via Civilization, a strategy-based game popular in the 1990s where players tried to build an empire that stood the test of time. 'I had to read a lot to understand what the game was about. To play the game well you need to understand history. Which technology gives birth to what technology, and which technology enables what,' he said. 'You need to have a marketplace, and then later you have a stock exchange. The technology tree.'

To beat the game, Wang discovered he could either conquer every other nation on the planet, or launch the first rocket ship from Earth. 'Of course I wanted to try both. That's always my approach,' he said.

Wang and his elder sister obtained undergraduate degrees in electronic engineering at Tsinghua University, China's equivalent of the Massachusetts Institute of Technology. They pursued PhDs in the US.

But Wang withdrew from his programme at the University of Delaware in 2003, sensing the opportunities kindled by social networks such as Friendster. 'When I saw the social network, I could instantly understand the importance and the potential of that.

'Social networking was going to change China. Someone had to build a social network for Chinese, so that's why I went back,' he said.

Enlisting his lower-bunk-mate in college Wang Huiwen and another middle-school fellow Lai Binqiang, the three raised 800,000 yuan from family and friends for their first startup, Xiaonei. Wang flew back to China on Christmas Day. He was twenty-five.

Wang set up shop near Tsinghua University to tap talent, running an office near the northern gate for cheaper rent. Tsinghua is anchored in the heart of Zhongguancun, better known as China's Silicon Valley. Once a mere state-backed technology centre, it took off after a Chinese academic proposed that the district be turned into a national high-tech park after a visit to the US in the 1980s. The place became famous as the 'electronics avenue' for peddlers touting pirated video compact discs, knockoff MP3 players and headphones. By the time Wang set foot on its twisting lanes, the neighbourhood had morphed into a pastiche of Mao-era-styled commune buildings and steely skyscrapers. A mini-Korean town sprang up within the locale, thanks to the thousands of students flocking to China each year, bringing in their wake restaurants, cosmetic bazaars and hip hop clubs that drew raucous crowds on the weekends.

The three of them knew how to code, but had to learn everything from scratch when it came to building a social networking website. 'Later we knew we knew nothing,' said Wang. 'I was the CEO, but I didn't know how to manage people.'

Xiaonei drew a solid fan base among university students, but Wang struggled to monetise it. Soon they burnt through all their money and he was forced to sell that outfit in 2006.

After half a year, he built his second startup Fanfou, a Twitter equivalent, reckoning that Facebook's decision to add a feed service was 'the most important change in the history of the social network'.

Wang's Twitter-like site Fanfou was halted within two years due to government censorship. Just like Twitter is banned in China, authorities deemed his platform a threat to the Party's rule as information often spread fast without control on such platforms. Wang learned an important lesson.

'I wanted to do something that's least controversial, and that helps people eat better,' he said. 'I hope nobody is going to have a problem with this.'

When I asked Wang what he learned about censorship and the lessons he took away from dealing with the government, he became defensive and went on a spiel about how the food-delivery industry is perhaps the least taboo among all internet sectors.

'It's perfectly normal. You have to deal with the government in all countries. We chose to do something least controversial. So we talk to the government, we follow the rules, the laws. We create jobs,' said Wang. 'I don't think we have a particular problem with the government.'[1] Wang's cautious tone at the time underscored the apprehensions that all Chinese private entrepreneurs harbour. The country's business landscape is littered with tales of billionaires who became just a tad too complacent, too smug, too vocal, too flamboyant, too political, and succumbed to imprisonment or even inexplicable deaths. The best way to describe how the government treats its internet tycoons is the carrot-and-stick approach.

---

[1] In retrospect, Wang may have been a little over-confident. During the internet crackdown of 2021, the same one that torpedoed Ant's IPO and thrust Alibaba under a regulatory microscope, Meituan and other gig-economy giants came under scrutiny as well for their treatment of often poorly paid contract delivery personnel. Meituan itself was subjected to a months-long antitrust investigation that ended only after the company agreed to cease certain non-competitive practices and pay a fine. The incident was a pointed reminder that no one is above the law – not even one that purports to just deliver food faster and more cheaply.

Even before the 2021 crackdown, under Xi Jinping, online restraints had grown tighter, particularly around the time of politically sensitive events such as the Communist Party Congress or anniversaries like the death of Nobel Peace Prize winner Liu Xiaobo. At such times, China steps up its effort to block virtual private networks, a commonly used method to circumvent the Great Firewall. Foreign companies that want to operate on the mainland are forced to adopt practices often seen as invasive elsewhere. Apple, which publicly fought requests by the US government to create backdoors into its password-protected products, has quietly deleted apps and built local data centres in line with Chinese government requirements. Microsoft said it would stop operating its work-oriented social network LinkedIn in China by the end of 2021, and Yahoo followed not long after by saying it would exit the country as well. All this contributes to China having the least online freedom on the planet, according to rights group Freedom House.

After a brief stint wooing the country's rising tech elite in 2015 – Xi named them as a key focus for the ruling Party's outreach, elevating China's internet parvenus to an unprecedented level of strategic importance – the country has all but cracked down on this new class of people.

Wang might still be mostly right, in that he's occupying one of the less touchy corners of the internet. Yet the vast amount of data his company sits on, the location-based consumption patterns of its 690 million active users – double the population of the US – is of value to the government. It's why Meituan wasn't spared when regulators went after internet firms in 2021. China's cyber laws demand absolute fealty from companies, requiring them to hand in data and cooperate with investigations if wrongdoing is suspected. Companies essentially are powerless in front of law enforcement when it comes to user privacy protections.

A popular joke among Chinese internet users: if you order up too many deliveries, especially during sensitive political summits, you might get the cops knocking on the door on suspicion that you're harbouring fugitives at home.

GALACTIC MANAGING DIRECTORS

Still, Wang thought little of Beijing's unseen hand during his early years of entrepreneurship. His main preoccupation back then was simply carving out a slice of China's booming online scene for himself.

In 2010, Alibaba was taking China by storm, with the internet population hitting nearly half a billion. After his first two failures, Wang thought instead of creating an e-commerce platform for products. He would build one for services that would become 'equally big or potentially bigger'. Meituan was born.

It was a heady time, replete with aggressive players looking to replicate Alibaba's success in different spheres. One of them was Rocket Internet co-founder Oliver Samwer and his golden child Groupon, the online discounts platform that at the turn of the last decade was one of the world's most-hyped startups. And then there was Tencent, teaming up with Rocket Internet in one of its first attempts to wage war on Alibaba via proxy.

Groupon and Tencent joined forces to roll out Groupon in the country in February 2011. They pledged $60 million, ferrying in US MBAs, Oxford and Cambridge graduates, bankers, consultants. The entrepreneur-aspiring type: young, mid-to-late twenties, foreign and always ambitious, according to Dave Chang, a well-toned and suave regional manager for Groupon China.

'They give you a really fancy title,' said Dave. 'The joke at Rocket was everyone was a managing director, a global managing director, then the really important people were Galactic Managing Directors.'

Setting his foot in Asia for the first time, Dave represented a breed of drifter expats that stumbled upon the opportunity of a lifetime. Part of his job was to help beef up Groupon's workforce in China in the blink of an eye.

'It was a mad scaling grindhouse. Are you a human being, do you have a pulse, can you do simple arithmetic in your head? Great. You have a job now. That's how you do it. That's how you hire 4,500 people in six months,' said Dave.

But he was outgunned – even with Tencent's backing.

Groupon's basic business model was to sell deals and discount coupons to customers and earn revenue as commission for every customer referred to the merchant. The operation was easy enough to replicate and soon copycats sprouted in about fifty cities in China within half a year, handing out subsidies indiscriminately to win market share. What it didn't anticipate was that 5,000 other so-called group-buying outlets similar to Groupon mushroomed across the country, with venture capital funds like GSR Ventures and CDH Investment piling billions into what became the two market leaders, Lashou and Wowotuan.

'Everyone was trying to outspend others, and it was a race to the bottom,' said Dave. 'It was the first time in Oliver's life where he couldn't outspend someone.'

IRON ARMY

Into the fray stepped Meituan.

Wang initially focused on food and dining. To compete with the scores of other group-buying sites, Meituan raised money from Sequoia in September 2010 and went on a hiring spree, jumping ten times in staff numbers to 2,000 people within a year.

His rivals fought dirty. Competitors began poaching his team by doubling salaries and elevating their titles. Many in the industry inflated transaction numbers, valuations, the amount of funds raised, and the rate of discounts offered. Wang disclosed his company's bank deposit balance at a press conference in July 2011, proving he really had raised $50 million in a series B round, as a means to protest against fraudulent figures.

For the first three quarters, Meituan fell behind and scrambled to keep up with its own growth – for a while, it wasn't even able to tally its staff in real time.

Wang was in real trouble. He had no operational experience. He had never managed more than twenty people. And within a year, his staff had grown to more than a couple of thousand. He needed an operational shakeup to stay in the race.

At the time, Alibaba executive Gan Jiawei was helping the e-commerce giant conduct due diligence on Meituan. Wang flew all the way to Alibaba's headquarters and the two had a cordial meeting in May 2011. 'The first time, he didn't leave a big impression,' said Gan. 'My first thought was he had a big forehead.'

Moon-faced and soft-spoken, Gan was an old hand at Alibaba, where he'd worked since the company's commencement. Colleagues and friends call him 'A Gan', as Forrest Gump is known in China.

A former staffer at a state-owned coal-mining enterprise, Gan ditched the comforts of the iron rice bowl and begged Alibaba to take him, sensing the opportunities to come in 1999. He rose through the ranks and led Alibaba's sales force, known as the 'Zhong Gong Tie Jun' (the central supply iron army). Even today, the name evokes awe and exudes an almost mystic aura.

Wang sat in Gan's office for hours lamenting how competitors were poaching Meituan's salesforce and wanted to know how he could deal with it. Those were the early days of the mobile internet in China, where companies relied heavily on ground forces to sign up merchants and convince users to download their apps. Competitors would often double or triple staff salary.

'Wang Xing's founding team, they were a bunch of Xue Ba (super good at studying) from Peking University and Tsinghua. In the past they never managed big teams,' said Gan. 'People who were adamant followers of Wang Xing believed in his vision and technology, but in daily operation, the company had no idea of how to manage its sales force.

'I told him this can't be rushed. Even if you poach the staff back, it won't work in the long run.

'If you cut the flesh off other people and stick it to yourself, it doesn't grow on you and will just rot, because it's not organic. You need to nurture your own team.'

One thing moved Gan during the meeting. Wang convinced him that Alibaba was a thing that represented the past – a platform for products – and what Meituan was trying to build in the long run was a platform for services, a much bigger addressable market that represented the future.

'I thought Meituan was focusing on a bigger market, bigger opportunities and faster development.'

Impressed by Wang's vision, Gan ended up joining Meituan in November that year as chief operating officer. He worked until 15 November at Alibaba, stepped out of his office in Hangzhou and took the first flight to Beijing the next day. 'I worked like crazy for twelve years. No break. And directly joined Meituan.'

In the later years, Wang Xing's arch-nemesis Alibaba would look back and conclude that poaching Gan was one the best things the Meituan founder had ever done. The veteran operator would prove instrumental in overhauling Meituan – not just introducing adult supervision but also streamlining decision-making, thus putting in place the systems it needed to knock Alibaba onto a back foot. Gan's changes, which have never been documented in detail, ultimately helped propel Meituan to the vanguard of the nascent meal group buying. Meituan would end up catching Pony Ma's and Tencent's attention in the process.

His tenure at Meituan started on a shaky note. Wang arrived in the office around 9 a.m. on 16 November, then came down one floor to greet Gan at his desk. He wasn't very articulate, and was even a bit shy when he introduced Gan to his two co-founders Wang Huiwen and Mu Rongjun. They took a photo together, which got lost through the years. 'It just shows you how badly we archived our own history,' said Gan.

Gan spent the next few weeks talking to Meituan's thousand-strong sales force. He essentially took charge of all the labour-intensive units. But the company remained confused about how to position its sales team and make it coordinate with other units.

'Management structure, commission structure, these things were non-existent,' said Gan. 'You can imagine it was like a peasant uprising. Yesterday they were farming, then overnight they became soldiers with guns. They didn't know the basic concept of clocking in or corporate governance.'

Meituan had a central command unit that micromanaged how many days a group-buying deal could exist on the website and at what discount the deals would be offered. It was quality control, but it also deprived salespeople of power. 'The front and the back office were not working in synch. A lot of infighting,' said Gan.

Gan said he banned salespeople from getting kickbacks from merchants, linked pay to sales performance, added regional managers to clear bottlenecks. Structure-wise, Meituan had seventy-seven city managers, two big regional directors and one vice president. That clogged reporting lines. 'Structure is hardware. You need your GPU, CPU to work,' said Gan.

Gan divided the country into eight districts and flattened the structure, giving more power to the district managers. He also elevated certain city managers to the same level of importance as a regional ones. 'Cities like Beijing and Shanghai were at the heart of the war zone; we had to react faster and give them better support,' he said.

In between all this, he also unearthed a critical insight: more listings lead to jumps in transactions.

It seems intuitive now. Yet back then most group-buying sites placed an expiration date on their coupons. In Meituan's headquarters, salespeople prioritised something known as 'consumption sense', which meant the best salespeople were able to spot what young people liked and would become trendy the next season.

Gan would have none of this. 'Consumption sense was something that couldn't be standardised or quantified. How can salespeople in a first-tier city say that they know what will become popular in the smaller cities in China?'

He figured that only by listing everything and anything, for as long as possible, could they win. Gan took his insight about how more listings would lead to more transactions to Wang Xing, who surprisingly responded: 'Of course, I knew that.'

'If Wang knew that at the time, it certainly was not apparent to the thousands of people working for him. The people on the front lines had no idea what they were doing,' said Gan.

The episode highlighted a flaw in Wang Xing's personality. While obsession with logic has won him many battles, his innate inability, or for that matter a lack of will, to communicate with people has been the source of many grievances in his relationships. 'Wang Xing thinks that he's successful because of his logical train of thought; he doesn't fully realise how important people are; he lacks an appreciation for people,' one of his former colleagues once told me. Even after his company went public, the billionaire admits out of frankness that he's not a good manager.

A huge debate ensued on whether the company should eliminate the policy of placing group-buying deals for only a few days – a way to boost their scarcity appeal. The team even sought out a famous temple in the west of Beijing for a retreat to dwell on the issue. In the end, they decided to free up supply.

Gan started granting salespeople the right to feature merchants as soon as they signed up. The company also introduced a refund policy for expired coupons. Transactions spiked soon after.

Then there was the competition. Meituan and rivals alike deployed teams to monitor each other. It was quite common for staff to work from 7:30 a.m. into the wee hours of 2 or 3 a.m., with three people doing the load of five. It was live or die at the time..

'Our employees were working much harder than other competitors. We were helping merchants launch their products much faster on the website. Sales could directly list merchants on our platform,' said Gan, who kept telling his staff: 'If you can survive at Meituan, your competitors will poach you for several times your worth.'

Meituan also set up morning and evening conferences so staff could share successful experiences. The company stipulated that salespeople had to come to the office at around 8:30 or 9 a.m. for a morning meeting and then again in the evening to talk with their team leaders every day. The sales reps would share what worked and what didn't with their unit heads, who would then spread the gospel to teammates the next day. This proved a stellar strategy to motivate members and train new hires.

The sales force did not like it, at all. Many pushed back, especially the divisions that were excelling. Gan was pretty flexible in execution – units that did well didn't have to comply, but if they were sub-par, they had to. Before long, the struggling brigades surpassed the ones that didn't follow this strategy.

What also differentiated Meituan was that Wang insisted the company focus on food and dining from the start. Many competitors expanded into selling merchandise such as chairs or hair-dryers. Wang reasoned it was important to have a high-frequency category to acquire and keep users.

By 2012, Meituan broke even. Competitors started losing momentum. Groupon's China business was folded into another entity backed by Tencent. 'When other people lost momentum, we kept going,' says Wang.

Tencent-backed Groupon lost out in the fight for group buying, partly because it didn't have a lieutenant like Gan. Tencent being a social media company didn't have the genetic code required to build a sales army, which is essentially what group buying is all about. Its strength lay in leveraging its instant messaging platforms to direct traffic to other content sites such as news portals and gaming sites. The group buying business was a completely different animal.

By this time, though, Tencent had also begun shifting its strategy towards investing in startups instead of trying to compete in every sector itself. Pony wasn't about to give up on getting into the so-called online-to-offline space, which focused on connecting people with real-world services via the internet. In fact, he saw this trend as the future.

ULTIMATE CONNECTOR

In late November 2013, Pony brought up a concept known as Internet + that would be elevated by the government to a national strategic level within two years. The idea was you could topple and revolutionise any industry if you linked it with the internet. Link the web with retail and you get e-commerce, with entertainment and you get online gaming. But there could be a lot more sectors that could be transformed, such as transportation, logistics, manufacturing and – most immediately – neighbourhood services.

For Pony, China's internet realm was on the brink of another transition. If web 1.0 was about connecting people with information in the virtual world, then the advent of the smartphone brought about chances to connect people with services in the physical world, something that never really took off during the desktop era. His vision for Tencent was to become the ultimate connector – linking people with people, devices with devices, people with devices, people with services – he knew that Tencent wouldn't be able to go at every sector and business itself, so the better strategy would be to back other companies, while it provided the underlying infrastructure via its platforms, initially QQ and later WeChat.

'Tencent can become an internet connector, on one end connecting its business partners, and the other its vast sea of users,' Pony said in an open letter in 2014. 'Tencent can connect everything.'

The strategy of being a connector is still very much the guiding principle at Tencent today.

Halfway across the world, Facebook's Mark Zuckerberg sketched out a similarly overarching goal for his company: connecting the world through social networking. The two companies had many similarities in strategy, in that they both wanted to get as many people as possible onto their platforms – and then get them to stay for as long as possible.

Where Tencent would outperform and even surpass Facebook was linking users to real world services more effectively. That's where its investment prowess and strategy of backing emerging startups came into play.

Meituan, at the time, was still squarely in Alibaba's camp. Tencent turned its attention to the other group buying site then vying for supremacy in China, known as Dianping. Founded by Wharton alumnus Zhang Tao in 2003, Dianping had corralled some of the biggest funds into its camp as well, including Google and Sequoia. That was to be Tencent's entree into the intensely competitive space.

To show their sincerity, Pony and Martin met with Zhang in person. They wanted to alleviate his concerns. 'I wasn't sure just how open Tencent really was. Tencent used to do everything itself, if that was still the case, then there would be no foundation for collaboration,' said Zhang.[2]

Sensing his reservations, Pony and Martin didn't cut straight to business during their first rendezvous. They spent a great deal of time talking

---

[2] 丹.声.道. (2014a). 大众点评张涛: 马化腾想明白了, 很脏很累很苦的活腾讯其实不愿意做. 虎嗅网. https://www.huxiu.com/article/38149.html

about the potential synergy the two companies could create by joining hands. One key message they got through was Tencent was painfully aware it was not capable of building its own online-to-offline operation. It was too labour intensive for Tencent, which was more accustomed to scaling fast via its online platform.

'Pony finally had a clarity of mind,'[3] said Zhang, joking that the group buying business would involve too much dirty and hard work. Dianping subsequently agreed to take Tencent's investment, selling it a 20 per cent stake.

But Meituan was rising fast, and Tencent knew it would someday have to be reckoned with. As fate would have it, Wang's company would end up on Pony's side – but it wouldn't be for a while yet, and not before a few detours along the way.

WAR ON FLIP-FLOPS

With Meituan making headway in group buying, the company was ready to take the next logical step.

Wang Xing dispatched his right-hand man, Wang Huiwen, to explore new initiatives. It took him a year – toying with every idea from Wi-Fi for diners to restaurant supply chain management. Meituan dedicated a team to reviewing rival apps that generated daily orders in the thousands. Then, one day, they spotted a gem: food delivery.

It was no blue ocean. Four-year-old startup Ele.me already blanketed main universities in the larger cities. The founder, Zhang Xuhao, was an avid gamer who'd graduated from a prestigious university in Shanghai. Early VC backers were impressed by his 'ugly feet', callused by all the food deliveries he made wearing flip-flops.

Wang Huiwen assembled a group of thirty people to set up food-delivery services in thirty cities, outnumbering Ele.me's twelve. The truth, though, was that it was a skeleton crew. It only ran operations on one

---

[3] 丹.声.道. (2014a). 大众点评张涛: 马化腾想明白了, 很脏很累很苦的活腾讯其实不愿意做. 虎嗅网. https://www.huxiu.com/article/38149.html

university campus per city. Meituan poached about five experienced hires from Ele.me, but most of the other people were clueless as to how to scale a logistically challenging operation.

They had a bumpy start in other ways. Wang Xing's business intelligence unit picked up on how Capital Today's Kathy Xu was talking to key figures in the food-delivery industry. He called Xu from the US, right when his wife was giving birth, to forestall Xu from investing in Ele.me and lobby for Meituan.

A no-nonsense, quicksilver conversationalist, Xu is one of the few successful female venture capitalists in China, known for backing e-commerce giant JD.com and Chinese travel site Trip.com when they were inchoate.

Xu, however, didn't invest in Meituan at the time – she was already a key investor in Tencent-backed Dianping. Wang Huiwen did manage to successfully sabotage Ele.me's fundraising to a certain extent. 'I thought Meituan would be a formidable competitor, so I'd better not invest in its rival,' said Xu.

Indeed, Meituan caught on quickly. Food delivery ran counterintuitively to almost everything about group buying. A typical example: group buying requires the company to list as many merchants as possible; for food delivery, reliable and fast service was key. One Meituan city manager dominated business in a university with a mere eight restaurants on the menu.

By April 2014, only two city managers had figured out this secret. Meituan summoned them back to headquarters to share their experience. Other city managers replicated their modus operandi. They also weren't above playing dirty. Ele.me posted ads on campus boards at 8 a.m. every day, so Meituan's staff would go at 10 a.m. and plaster their own posters on top.

For the merchants who agreed to exclusive deals with Ele.me, Meituan's staff said they would sign them up for delivery without the merchant's consent, and offer its own fleet to help ferry food. Because of the significant web traffic it brought in, merchants were happy to work with them and renege on the exclusive contracts. Within a year, orders started picking up.

Meituan now had a decision to make: whether it wanted to bulk up and run its own fleet of delivery people to guarantee service quality. It would be a huge change – taking on a much more operations-heavy and margin-eating operation. But 'it was the only logical thing to do,' says Wang Huiwen. The company wanted to minimise its reliance on restaurants for delivery.

It did so by building a hybrid self-owned and franchise model, where third-party companies could take over and run a delivery station that covered a two-square-kilometre radius. As of 2019, most of Meituan's 700,000 delivery persons were franchised.

Wang Xing says they won with reliable technology that helps optimise routes and track riders in real time. He had few issues about copying other companies, labelling Facebook a copycat as well. By his logic, only the businesses with the best services can win because that's what customers care about. Not innovation.

So it was in 2014 that Meituan was fighting at least two major wars – group buying and food delivery – while waging smaller battles, such as in movie ticketing.

Three serious contenders were left in group buying: Dianping, backed by Tencent; China's search engine Baidu; and Meituan, backed by Alibaba.

EPIC COUP

Meituan and Dianping were together burning about $2 billion a year, estimates Bao Fan, founder of boutique investment bank China Renaissance, which was an investor and financial adviser to both firms.

The punctilious Wall Street veteran honed his rapid-fire English while working at Credit Suisse and Morgan Stanley. His boutique investment bank was famous for being able to broker unthinkable mergers, while taking fees from both sides. Compact and taut, Bao Fan commanded respect and deference. Based on his deal record, he was entitled to it.

The cash burn for subsidies that so worried Bao was getting unsustainable. In order to get restaurants onto their platform, the three main players had to underwrite sales, paying caterers millions upfront. 'The burn rate was big and the cashflow was diminishing every month; it was scary,' agreed Xu, the Capital Today founder who was a backer and board member at Dianping. 'You needed a strong heart.'

Or, an ally.

Meituan and Dianping had flirted with the idea of a merger as early as 2013. It never went anywhere because money was ample and no clear leader had emerged. Now that the spigot was drying up, people started taking a more serious look. 'There were questions of sustainability,' said Bao. 'Fundraising became difficult for both companies.'

That was when Meituan's relationship with Alibaba got complicated – and Tencent went in for a kill.

Alibaba throughout the years tried to fully integrate Meituan – not just angling to put a search tool on Alibaba sites, but also to build Meituan's services database, including its valuable merchant and customer data, directly into Alibaba's shopping site, according to co-founder Wang Hui-wen. Meituan refused.

Exacerbating the rift, Wang Xing sometimes simply ignored requests from top executives, according to people familiar with the situation at the time, giving them the impression that he was arrogant and uncooperative.

When Meituan raised $700 million in venture funding around early 2015, Alibaba requested to increase its investment stake to about one-third of Meituan, so the two would have better synergy. Wang, who preferred to maintain his independence, promptly refused.

'We can all agree that Alibaba was a very ambitious company, and they kind of have a weird way of thinking that "we own all commerce. If you are doing anything related to commerce, you are stealing money from me,"' said Wang.

Sensing the wars to come, Meituan started raising another round of funding in July. This time Alibaba said no, having committed $1 billion of investment into its own local services site, Koubei, in June.

Now, Wang was in real trouble. On the same day that he landed in San Francisco to raise funds, Chinese search giant Baidu announced the company would devote more than $3 billion to building its presence in the services industry. 'It made it much more challenging for other financial investors,' said Wang.

Wang faced the quandary of either working with Baidu or swallowing a down-round by raising at a lower valuation.

Then a third option emerged. Capital Today's Xu caught wind that Meituan, like Dianping, was struggling to raise funds. She passed the tip on to Martin Lau, president of Tencent, backer of Dianping, nudging him to broker a merger between the two. 'No merger, no money,' said Xu. It would be very difficult to strike a deal if they just let the two founders talk it out themselves; 'We needed a white knight and that was Tencent, so in stepped Martin.' He met with Meituan's Wang Xing via a meeting brokered by his former underling Chen Shaohui, who had then become Meituan's chief strategy officer.

The two met in the eastern financial district of Chaoyang at a Japanese restaurant. Tencent was willing to lead Meituan's fundraising by pledging $1 billion, bringing in its associates and partners after a consolidation and letting the company operate independently. 'It was a very easy meeting. What they had, we needed. What we had, they needed,' said Wang.

All in all, it took ten days to hammer out a deal. The team worked in stealth mode. The merger was codenamed 'Race'. Dianping was referred to as Ducati, Meituan was Maybach. Negotiations continued in Hong Kong, where discussions lasted for ten hours, well beyond midnight, and no one was allowed to leave the negotiation room. They even decamped to Australia when Martin and two other key Meituan backers, Sequoia Capital China founder Neil Shen and Renaissance Capital CEO Bao Fan, attended the wedding of Richard Liu, the CEO of e-commerce site JD.com.

'We were kind of moonlighting a bit at the wedding, but we had no choice because everything had to be under stealth mode,' said Bao Fan.

When Meituan called a board meeting to make it official and announce it had taken investment from Tencent, Alibaba got twelve hours' notice and no choice in the matter even though it had veto rights on investments, according to people attending the meeting. The reason for this was because Meituan structured the merger based on Cayman Island laws that green-lit deals as long as they obtained the requisite board and shareholder approvals. Theoretically, Alibaba could have sued Meituan, but it didn't.

Jack Ma was furious. He saw betrayal. Tencent saw victory. Wang saw survival.

Baidu, which toyed with the idea of backing Meituan, had been too hesitant and stingy about valuations – a credo that has caused the search engine to lose out on some of the most important acquisitions throughout the years. Baidu's neighbourhood on-demand service – the business it pledged $3 billion toward – became a shadow of its former self and was later sold to operations that became part of Alibaba.

The whole episode left a bad taste with Alibaba. The e-commerce giant later tried to mess with Meituan by dumping the company's shares at a $12.5-billion valuation, a 40 per cent discount to what Meituan was seeking in yet another post-merger financing round. Alibaba declined to comment. It made fundraising extremely tough.

But Tencent, now fully on board, proved instrumental. In the end, Yuri Milner's DST and Singaporean sovereign wealth fund GIC came to the table.

'The group buying wars set the template for the cash-burning model for the ride-hailing, bike-renting wars to come. It became a competition template in China for any platform-driven, winners-take-all game,' said Bao Fan.

One of the trickiest issues during the merger was which of the two company founders would remain CEO. Wang was adamant from the start that he wanted to run the show. There were a few things working in his favour: Meituan's market share was double that of Dianping's, bolstered by early expansion into smaller cities; Meituan also boasted a two-year-old food-delivery business, which Wang touted as a growth engine.

In the end, the pair agreed it would be a merger among equals, with the understanding that Dianping's founder Zhang Tao would someday hand over the company to the younger Wang Xing – Zhang was born in 1972, and in China that seemed almost geriatric for a startup founder.

While Silicon Valley has been criticised for its obsession with youth, China is emphatically unapologetic. The cut-off age for programmers and product managers often tops off at thirty, while sales staff can be limited to the ages of twenty-five and below. China's largest job-recruitment platforms are littered with thousands of job ads that blatantly place caps on candidates' ages.

In truth, Zhang had always looked forward to retiring at some point after a bruising five-year battle. But his departure came much sooner than expected.

The freshly minted entity faced a pressing dilemma immediately after the merger: whether to fold its food-delivery business into rival Ele.me, or whether to run the operation itself. Zhang's Dianping previously held a significant stake in Ele.me, and if they chose to fold, it would naturally give Zhang more power and upset the duo's dynamic. Wang Xing preferred they run the food-delivery business themselves.

Fierce debates ensued between the two. Wang Xing and his lieutenant, Wang Huiwen, spent a lot of time convincing Tencent's Martin to let them run the food-delivery business. In the end, Wang convinced Martin that only via direct operation could Meituan–Dianping control the service better.

The alternative would have been running a group-buying operation while owning a stake in a food delivery portfolio company, but not having direct execution power over the business. That would have caused a lot of issues, especially given that Alibaba and Baidu were both still in the fight at the time.

Whatever the logic, history shows they made the right decision. Meituan in ensuing years managed to gradually squeeze Ele.me out of certain strategic markets, expanding its market share and ultimately building the world's largest meal delivery service.

That years-ago decision is what still partly underpins Meituan's $200-billion valuation, while Groupon trades at only $670 million in comparison. It also unfortunately planted the seeds of a cut-throat battle between Alibaba and Meituan.

BATTLING ALIBABA

Zhang Tao left soon after that. Wang says Zhang lost support among investors. Zhang declined to comment on the reason. But to the latter's credit, he 'had a big heart, he was looking at the bigger picture,' said James Mi, founding partner of Lightspeed China Partners, an investor in Dianping.

Alibaba later bought Ele.me at a $9.5-billion enterprise valuation. Meituan also tried but failed to win the deal by offering $9 billion. Alibaba by now had also acquired Baidu's delivery business, and combined it with Ele.me and its own inhouse unit, Koubei. In August 2018, the enlarged business raised more than $3 billion from investors, including Japanese investment powerhouse SoftBank. Alibaba spent about $500 million in the third quarter of that year alone on subsidies and discounts for its food-delivery platform.

Yet with Tencent's backing, Meituan held up under the fusillade. That doesn't mean Wang holds anything back when it comes to Alibaba. He laments how Jack Ma impaired the reputation of Chinese entrepreneurs when he plucked Alipay out of the e-commerce entity without shareholder consent. 'I still think he has an integrity problem,' said Wang. 'I think the impact of that incident is still underestimated.'

Taking a page from Jeff Bezos, Wang remains in no rush to become profitable. He's prepared to reinvest in new business as long as the company has a healthy cash flow.

The entrepreneur's meteoric ascent provides a unique lens through which the outside world can better understand Tencent.

Tencent's investment into Meituan in 2015 encapsulates the social media giant's favoured approach over the years – backing the right horses, then taking a hands-off approach.

Pony and Martin recognised in Wang Xing a sort of proto-Pony, an entrepreneur with the vision and unswerving will to take on new arenas and competitors, swiftly and on a grand scale, adjusting course when necessary but always preferring autonomy.

Ironically, one of Meituan's most important overhauls – of the sales force and decision-making processes – was overseen by an old Alibaba hand, Gan.

Tencent's dealings with Wang would set the template for future competition with Alibaba over and over again throughout the years, including in payments and ride-hailing, which would involve not just Meituan but another major Tencent investment, Didi Chuxing (now Didi Global), the startup perhaps best known for defeating Uber in China.

But to get into that story, we need to begin with Tencent's most powerful creation, WeChat.

CHAPTER 7

WECHAT, CHINA'S VIRTUAL TOWN SQUARE

Imagine for a second living without a phone. You could get by – barely. Simple luxuries like ordering takeout, hitting up friends and family, and buying groceries become almost impossible, never mind dealing with emergencies. In China, that's sort of what it's like to live without WeChat – the online forum, town hall and marketplace for a billion people. Despite its name, WeChat is much more than a messaging service.

When I visited WeChat's headquarters, it had already become a super-app that hooked up a billion users, three times the population of the US. Today, WeChat is the largest social media platform in the world after Facebook and WhatsApp, attracting more users than Twitter and Snapchat.

Inside WeChat's campus in the balmy southern city of Guangzhou, creepers snake along brick bungalows shaded by lustrous camphor. But for the incessant songs of cicadas, it's a quiet industry park delineated by unmanned, automated convenience stores and chic cafes. Programmers hunch over their open desks in retrofitted communist-era cotton factories, collectively contributing to a product that weaves together the fabric of society in China. Think of it almost as the operating system for all basic online activities in China, from gaming and shopping to getting a ride and booking a doctor's appointment.

As an experiment, I clocked how long I could stay on the app while running around town. After messaging my friends on WeChat, they sent me the location of a restaurant at which to meet. I hailed a cab, listened to Taylor Swift, booked movie tickets for *Spider-Man* on the way, and then paid the taxi driver. Upon arriving at the diner, I scanned a QR code and was redirected straight to a menu. We placed our orders and paid the bill, all the while hardly interacting with the waiter. On my way back, I booked flight tickets and a hotel for my next trip and scrolled through my friends' social media feeds, taking in the latest news and celebrity gossip of the day. All this time, I never left WeChat.

Sure, it's just an app – but an incredibly powerful one, an app to end all apps. You can call taxis, donate to charities, play games, and send and receive money. It's a hub for apps, services and customers. Nearly 99 per cent of internet users in China access it through a mobile. And they spend about a fifth of their time on Tencent's all-purpose platform. WeChat leads any other piece of software by a wide margin in almost every facet of online existence – in gaming, payments, social media, music and entertainment. Fundamentally designed as a messaging tool, it's come to mesh the functions of PayPal, Citibank, Facebook, Instagram, Spotify, Expedia, Yelp, Uber, TikTok, Amazon and Tinder – like Android, but directly overseen, nurtured and controlled by one giant company.

It's this incredible usefulness and convenience that has a billion people hooked – many every single day.

It's become so powerful and pervasive that it attracted the attention of President Donald Trump in 2020. In the run-up to the US election, Trump threatened to ban the app on the grounds that the Chinese Communist Party could use it to spread disinformation, censor news critical of China and steal users' private and proprietary data. That's despite WeChat hardly registering with most of the American public – its US users are estimated in the mere tens of millions, a drop in the ocean of its global audience. Yet Trump's rhetoric triggered hysteria in the market, slashing a tenth off Tencent's market value on the day. Tencent declined to comment on the matter at the time.

The episode – the ban got stalled by a court injunction and the Biden administration launched its own security review – could still be far from over, underscoring the influence WeChat commands.

More broadly, with the emergence of companies like Tencent, China's tightly controlled internet model has risen as an alternative that many emerging countries in Southeast Asia and Africa are closely observing and even aspiring to.

WeChat, however, didn't start out with the vision of being the one app to rule them all. Tencent and Pony Ma, having scored success after success in PC gaming, foresaw the mobile revolution and decided they needed a piece of the action. The WeChat origin story has by now taken on a sheen of fable and legend, capturing the imagination of a Chinese public hungry for heroes that can hold their own against the West. The story of a small but formidable team that managed to beat dozens of far larger competitors with sheer determination (and deeper pockets) is one that many Chinese are familiar with – yet it remains unknown to many outside of tech-industry circles.

WeChat's conception is also a story of compromise and controversy – how politically unaware tech geeks found their product in the crosshairs of the Chinese government, and through a gradual series of tweaks in response to periodic crackdowns and cycles of self-censorship, it became the go-to app for surveillance. During the Covid-19 outbreak in 2020, the government was quick to resort to the app for population control, assigning colour codes to determine whether they were healthy and eligible for things as basic as leaving their residences.

When you consider that the country's denizens are scattered across an area larger than that of the US, you can imagine the enormous utility to Beijing's apparatus of an app that virtually everyone uses on a daily basis. Some say WeChat should be called WeCheck instead. The app's social media function became an important avenue for people to vent their frustrations.

Tensions came to a boil in the spring of 2022, when a resurgent outbreak of the omicron variant – for a time – surged seemingly unchecked through the financial centre of Shanghai. Subsequent lockdowns confined millions to their homes, triggering mass-testing on a scale unseen since the initial outbreak and predictably outraging affluent urban residents who thought they could count on the government to safegaurd their wellbeing.

Beijing grew worried about the mounting discontent surrounding China's Covid-zero formula of lockdowns and widespread testing to wipe out the virus. To regain control of the narrative, the government ordered internet platforms to wipe and scrape posts deemed negative or critical of the policy, sending WeChat censors into overdrive. But it also further antagonized people whose lives had once again been up-ended.

Those frustrations culminated in an unprecedented public outcry on WeChat in April 2022, in what essentially became a virtual protest. As Shanghai went into lockdown, a video documenting the dire fallout began circulating online. The six-minute clip known as "Voices of April" - a montage of audio recordings encompassing the cries of babies separated from parents during quarantine, residents demanding food and the pleas of a son seeking medical help for his critically ill father - resonated with the tens of millions in Shanghai and more across the country. The video was quickly marked as banned content and taken down from social media platforms in China. On the Twitter-equivalent Weibo, even the word "April" was temporarily restricted from search results.

Netizens were infuriated. Many deemed the video a neutral yet essential documentation of the human toll of Shanghai's lockdown. A backlash ensued, as defiant users repeatedly shared the video in ways that could dodge web censors. Some posted the video upside down, others superimposed words or images or embedded other footage. WeChat censors tried to wipe posts sharing the video, but it was like a multi-headed hydra -- no sooner when one got blocked, another would pop up.

It was a seminal moment.

"I feel like I'm witnessing a historic event," one of my contacts posted. "I can feel how sad and enraged people are about their posts getting deleted from Moments," another said. "404 is just anaesthesia, not a cure. People will develop resistance to anesthetics," said another contact, who typically shies away from anything remotely sensitive (404 is the universal message that pops up when a site is inaccessible or no longer exists).

WeChat's "Moments" function had up to that point become a mash of typical Instagram and Facebook posts, where people shared their finest moments dining at Michelin-star restaurants, admiring Beijing's pink-hued clouds at dusk, or brooding on epithets of wisdom. As I scrolled through my feed that Friday night in late April, the platform became a cascading waterfall of images and texts revealing the raw emotion and anger of people – a rare moment where Tencent's signature service amplified the hurt and pain of a collective psyche.

ORIGINS

WeChat's origins were innocent enough. In 2010 the world was holding its breath for Apple's iPhone 4. Telecom operators across the world were preparing the switch from 2G to 3G, which promised far faster mobile internet connections. Tencent controlled a large swathe of social media and messaging via its QQ desktop app. Yet Pony, already a billionaire thanks to Tencent's early forays into PC and online games, was concerned his empire could get overturned overnight by some new mobile startup. Instead of succumbing to another competitor, he decided it made more sense to self-disrupt – to dream up a product with the potential to lay waste to Tencent's core business. 'It was a matter of life and death. Speed determined whether our company could survive,' Pony said.[1]

Within his own company, three teams set out to find an answer to mobile messaging. On a very late night towards the end of 2010, a programmer named Allen Zhang pinged Pony asking for permission to develop a social network tailored for smartphones. Pony, who usually doesn't go to bed until 4 a.m., agreed.

'It's hard to imagine how central WeChat has become to Tencent,' Martin said. 'We knew that there was user need for a mobile-only instant-messaging product that was homegrown. 'The degree of its success is not something that Tencent foresaw.'

In retrospect, Allen Zhang – who would go on to create the now-ubiquitous app – was one of Pony's best hiring decisions. A reticent geek with hair parted all the way from his right ear and combed across a huge dome of a forehead, he often speaks haltingly, as if scared of the words that escape his own mouth. The WeChat founder still today shuns the limelight

---

[1] HKU media file.

but inspires a cult-like following. When he does show up in public, he commands a singular reception. In a surprise appearance in 2019, he managed to stretch an originally thirty-minute-long speech into a four-hour soliloquy that kept an audience of thousands glued to their seats till nearly midnight.

SHARK WOMB

Allen was born on 3 December 1969 in the throes of the Cultural Revolution, which struck hardest in regions such as his hometown of Hunan, birthplace and revolutionary base for Mao Zedong. Perhaps that's why Allen has been cautious much of his life about broaching his political views. A former colleague described the entrepreneur as being 'extremely prudent when it comes to politics'.

Instead of making it into the top echelons of universities, Allen stayed closer to home and studied electronics engineering in the neighbouring province of Hubei. Shy and introverted, he occupied himself by hanging out in the computer labs of the Huazhong University of Science and Technology. He found solace in the books of Friedrich Hayek, best known for the writer's defence of liberalism. Due to his struggles with socialising, Allen spent the greater part of his life seeking the answer to communicating with people better.

Aspiring to be part of something great, he rejected a cushy job opportunity at a state-backed telecoms company upon graduation in 1994. The twenty-four-year-old took one look at the oppressive building and this isn't how he wanted to spend the rest of his life. So instead he headed south, as far as he could go, all the way to Guangzhou.

To Allen, the southern coastal metropolis represented opportunity. China's third-largest city, nearly ten times the size of New York, was in the midst of a manufacturing and trading boom. Home to the world's largest commercial drone maker, DJI, and network and smartphone giant Huawei, the city was fast earning a reputation for being China's very own Silicon Valley.

His first stint was less than successful. The startup he worked at went belly-up within two years. After finding a second gig, he decided to side-hustle, zooming in on building an email service. After work, Allen would focus on his passion project, researching, designing and putting together 70,000 lines of code, all on his own.

Allen offered his creation – Foxmail – for free. For a long time, he didn't know if anyone would pay for it. One of his internet compatriots, Zhou Hongyi, who later founded Qihoo, recalls that when he visited Allen in Guangzhou, he told the diffident programmer to monetise his product by introducing ads. Sitting in a cramped, smoke-filled office, Allen asked him why it was necessary to commercialise the service. To Allen, showing dedication and devotion to creating a good product was enough. He never charged money.

By 1999, Allen had quit his full-time job at a company that created document management software so he could focus on his now two-year-old passion project. He detested the constricted lifestyle of working at a corporation – a trait he harboured even after becoming a heavyweight at Tencent, often finding excuses to skip weekly executive meetings.

He was barely getting by, freelancing to make just enough for a living and maintain Foxmail. He had his doubts on whether the email product would ever be valuable – at one point he considered giving it away so he could pack his bags and go to Silicon Valley.

Instead, Allen came across some business opportunities. After his product attracted more than 2 million users, the twenty-nine-year-old Lei Jun – best known later in life for creating Chinese smartphone name Xiaomi – reached out and asked whether he was interested in selling. Allen timidly proposed 150,000 yuan – roughly $20,000 at the time. Lei, who was busy running his Kingsoft antivirus software firm, never responded. Luckily, two years later, a Chinese company called Bo Da acquired the email service – and Allen himself – for 12 million yuan.

Allen regretted that decision. 'I can recount every detail of Foxmail, inside and out …'[2] he wrote in a letter the night of the acquisition announcement. 'In my heart, it has a soul, because every line of its code

---

[2] Allen Zhang's study notes (2019)

contains the memory of my creation. I suddenly had an urge to renege on my decision.'

He bought a car and went off to Tibet when he turned thirty. His life wouldn't take another turn until five years later. In 2005, Tencent acquired Foxmail for $5 million, local media reported, citing people familiar with the deal. Allen joined Tencent, allowing him to continue to focus on his beloved mailing service. That expertise in email structures later proved instrumental in laying the foundations for WeChat.

Allen's team, however, wasn't the only one within Tencent working on a mobile instant-messaging app. Two others were racing to outdo each other, including one hived off from QQ, the division that contributed most to Tencent's success in the desktop era by catering to people's social media, gaming and news demands. It's not unusual for Tencent to allow multiple internal teams to compete with each other; it's in fact a trademark of the company's that ensures innovation.

'Tencent's culture is like a shark womb,' said Andy Mok, a senior research fellow at the Center for China and Globalization, alluding to how some unborn sharks cannibalise siblings in the womb to ensure their own survival. 'It's not as deadly, but it makes every member adapt faster and be more competitive.'

LOSING A $100-BILLION BUSINESS

Truth be told, WeChat was a dud at first.

Allen and ten of his colleagues sequestered themselves in an office covered with old-school whiteboards – the type teachers scribble on in classrooms. That team later became known as the 'little black-room 11'.

The group, comprised mostly of fresh graduates, was able to conjure up the first version of WeChat. 'We were under so much pressure at the time,' Chen Yuewei, an early WeChat hire, told China's state broadcaster. 'We managed to beat our competitors because of our better technology.' The reality behind that story was much more convoluted, and inconvenient.

WeChat launched its first version in November 2010. It had four functions only: profile photo and user name setup, simple texting, photo transfers, and contact list import. At the time, apart from directly signing up, one could register for WeChat using one's QQ account or email. Looking back at WeChat 1.0, it's evident it was conceived around how personal email contact lists are structured.

Even in its early days, it was criticised for bearing too much resemblance to the Canadian messaging app, Kik. One of the most iconic images of the mobile era is that of Allen speaking into a smartphone just like it was a walkie-talkie – unheard of for phone users more accustomed to finger-typing. It illustrated WeChat's popular push-to-talk function, which allows people to leave one-minute voice messages that show up as dialogue bubbles. What's not as widely known: that function is also a rip-off.

WeChat's first version struggled a great deal and gained very little traction, garnering only 1 million people within six months. In China's internet context, those figures were an absolute flop.

Part of that stemmed from Allen's own shortcomings – in his being unable to clearly communicate his vision. While known as one of the best product managers in China, Allen struggles to convey his demands or vision to those around him. 'It was absolutely painful working with Allen,' said a former early member of the WeChat team, who requested not to be named. 'We all had to constantly guess what he was trying to say, come up with seven or eight versions, and he would just shake his head and tell us it was not what he wanted.'

Another reason was that someone else had stolen his thunder. WeChat's most potent competitor at the time was revving up just across the border. In the financial hub of Hong Kong, a local startup was hatching an app known as TalkBox.

Text messages were expensive in Hong Kong around 2010, costing 13 cents each. WhatsApp, which was created a year earlier, was an instant hit among smartphone-savvy, price-conscious consumers. When the iPhone opened up push notifications for developers, it also opened the floodgates, spurring an explosion in messaging apps. For the older generation in Hong Kong and mainland China, WhatsApp was hard to use and many refused to type on small screens – an opportunity for smaller, locally savvy competitors.

Heatherm Huang, a twenty-one-year-old college graduate at the time, came up with an answer. Forthright and unvarnished, Huang affects a cyberpunk look with a goatee and buzzcut accentuating a long pony-tail. For weeks, Huang had been struggling to teach his own parents how to use WhatsApp on the iPhone. They simply refused to text.

Then one day, he had an epiphany observing Hong Kong's cab drivers talking to each other on walkie-talkies. He envisioned an app that would allow people to send messages by simply pressing a button and recording a minute of their conversation. 'Inputting Chinese was much harder than English,' said Huang. 'That's why the push-to-talk function was so popular in Asia.'

TalkBox went on the market in January 2011. Huang and his team thought they could imitate WhatsApp and charge people $1 to download after the product hit a million users. They estimated it would take about one year or two. It took them a month. Chinese internet guru and former Google China head Kai-Fu Lee lauded the product as one of the most interesting apps to watch. For months, it was the No. 1 downloaded app not just in Hong Kong, but also in mainland China.

Huang only truly understood his product had gone big-time when Chinese copycats came out of the woodwork. Late one night, he recognised a username that popped up just a few weeks after TalkBox's launch: Pony Ma. A week after that: Allen Zhang. 'I was so surprised Pony and Allen were testing our products themselves,' Huang said. 'It was quite surprising that as the heads of such a big company they were such avid product testers.'

They must have learned enough to squelch TalkBox in its infancy. Four months after TalkBox was introduced, WeChat came up with a copycat by incorporating the push-to-talk feature.

Mobile internet speed was relatively slow at the time – for a thirty-second message to go through, it would take roughly the same amount of time – a killjoy for increasingly impatient users. TalkBox devised a mechanism that would send the voice recording bit by bit even before users finished talking. 'It gave people the illusion that the product was faster,' said Huang. Other copycats only imitated TalkBox's interface, but didn't bother to incorporate this one vital feature. WeChat was the exception. 'That's when we knew they were taking our product apart and studying it very closely,' said Huang.

Then Huang received a call from Martin Lau. The Tencent president wanted to buy his company. The call offered a sense of validation but also dread. In China, startups dub acquisition bids from Tencent an offer they can't refuse. There was a lot of 'I buy you and kill you or I just plagiarise you and you die mentality', said Huang. TalkBox, wary of the circumstances – and blindly optimistic about its ability to duel with Tencent in China – decided to remain independent.

TalkBox made its first mistake a few months into launching. In 2011, most people were still relying on 2G in mainland China. The most popular data package in China was 30 megabytes for about 80 cents. The novice founders, thinking 3G would soon dawn upon the country, decided to opt for voice quality instead of speed. They were wrong; the faster network didn't roll out fully until 2013, missing their estimate by a year.

WeChat chose the reverse and soon started picking up momentum. A TalkBox message for one minute used up more than a tenth of one megabyte every time. With only thirty megabytes in their data packages, many users were reluctant to send voice messages using TalkBox. WeChat, on the other hand, only required a fifteenth of TalkBox's.

Herein lies an important lesson that wizened startup founders the world over swear by. Timing is everything – come up with an idea too late and you miss the wave, yet roll out something too early and people will deem your product useless. The same happened with food-delivery and restaurant group discounts during the desktop era. They never took off.

In a market where copyright infringement is rampant, product design and execution trumps original ideas. Some would argue that plagiarism throttles innovation and deters creators. Yet others argue it forces companies to come up with the best products based on micro-innovation. In any case, TalkBox never bothered to sue WeChat or Tencent, which would have been an extremely arduous endeavour with little prospect of success under China's legal system back then.

Huang's app stalled going into the second half of 2011 after WeChat borrowed his best features, leaving the Hong Kong startup in the dust. AI guru Kai-Fu Lee called it: 'The missed opportunity that lost a $100-billion business.'

WeChat reached 100 million users in 2012 and tripled that number a year later. It overtook not only TalkBox but even Tencent's own formidable QQ. The venerable desktop-messaging service's managers were so distracted by a battle with Weibo, China's Twitter equivalent, they hardly devoted any time or ammunition to its own mobile version. QQ's early phone-based iteration was extremely hard to use – it didn't incorporate push notifications, so people never knew when others texted them offline. On the other hand, WeChat tapped into QQ's contact lists, leveraging the latter's hundreds of millions of users. WeChat even notified people when they received messages on QQ.

The fact that QQ's own mobile version lost out to an internal underdog has become a classic business case-study on the power of unpredictable disruption. It wasn't as if Tencent set out to supersede its early hit – it's just that QQ lost sight of where the future was heading and missed out on the arrival of the mobile era.

'You either wait for someone else to kill you, or you kill yourself first,' is one of Pony's most often-quoted mantras. That value, which has been instilled into every employee, explains Tencent's paranoid leadership.

In hindsight, Huang now sees how ridiculous an idea it was that his startup could ever take on a goliath. TalkBox was operating outside of the mainland, so its servers were offshore, dragging down internet speed. It was also only a matter of time before the Chinese government, wary of non-domestic tech companies, would clamp down. TalkBox, along with Line and Kakao, were blocked in China a few years later.

A crackdown on WeChat, however, was also in the making.

CENSORSHIP AND SELF-CENSORSHIP

After WeChat surpassed 100 million users, it turned to what Tencent did best: building a social media platform. In April 2012, it added its most famous function, 'Moments', the closest equivalent to Facebook's social media feed, a stream of photos and articles that people can follow in real time.

Taking a page from QQ's playbook, Tencent's founders tried to find ways to build more services they could offer on WeChat from day one. Again, it saw an opportunity to cross-sell users content, news, and perhaps even music and games, one step at a time. It was a drastically different mentality compared with that of the WhatsApp founders, who doggedly pursued simplicity. 'Driving the strategic integration between products was something that we established really early on,' said Tencent's chief exploration officer, Wallerstein.

WeChat soon realised that in order for people to post on their social media page, the app needed to provide content. Within three months, it introduced a service known as 'Public Accounts' that would allow writers, journalists and government agencies to create the equivalent of blogs under registered public accounts. If people liked the content they could simply repost the articles and also follow the accounts. WeChat had evolved from a basic texting tool to a social media platform.

Once conversations seeped out of the private domain and into the public sphere, it elevated WeChat's stature – and also got it onto the radar of China's cyber watchdog The government stepped up its scrutiny, prompting Tencent to delete blogs, articles and suspending public accounts on WeChat.

Crackdowns came quarterly if not monthly. In one fell swoop, Tencent once deleted 85,000 'rumour' or fake news articles and penalised 7,000 accounts for violating regulations. Offences included everything from spreading false rumours and vulgar content, to information deemed politically sensitive or critical of the government.

The constant censorship hampered WeChat's popularity. It crushed people like Laura Lian. At the peak of her WeChat blogging career, Lian was earning about $7,000 a month, writing satirical articles for more than 220,000 fans. As a twenty-six-year-old Beijing resident who often crossed paths with the Chinese and expat community, Lian had a keen eye for spotting gaffes and irksome traits from both camps. Her account stood out for its blunt and edgy humour. She even won backing from an investor. Then WeChat shut down her blog with no warning. It happened just after she posted an article mocking Chinese men's hairstyles, including former President Jiang Zemin's slick-backed coiffure.

'I never thought it would be this serious because my article was just about his hairstyle and it was in English,' Lian said. 'I didn't realise I was never getting back this account and all my followers.'

Lian followed WeChat's instructions and filed a letter of explanation within the 100-word limit. A week later she got a curt response: 'not approved'. 'Even if you shut down my account, give me a clear reason,' Lian said. 'Don't give me a page full of regulations and laws. I don't even know which rule I violated.' She switched platforms after that. Tencent didn't respond to comments at the time.

It didn't take long for regulators to start tracking private conversations, and such surveillance soon ensnared a variety of users. In 2014, Chinese political activist Wen Yunchao was abruptly locked out of his personal account after setting up a chat group called 'Shorting China' to share information on protecting human rights in the country. Wen, a rights activist living in New York, said he got a notice saying his WeChat account was banned for serious violations of company policies.

When Wen tried to open his account, a message popped up saying, 'Your account has been closed and cannot be used to log in to WeChat', according to a screenshot he shared. After he applied to unblock his account, he got another message that read, 'This WeChat account has seriously violated our policies and is permanently banned.'

Wen was lucky that he had moved to the US. Others have been detained by police and sentenced for months for sending messages critical of the government in private chats. Beijing sent chills across the population of WeChat users in 2017, when it stipulated that administrators were personally liable for whatever was said on chat groups they run. Even before the announcement, Chinese authorities disciplined forty people from one WeChat group for distributing petition letters, while arresting a man who complained about police raids.

'If you are a group chat leader you have two choices: either you are going to super-actively monitor the group, because your livelihood is at stake, or you're going to delete the group,' said Lokman Tsui, author of *The Hyperlinked Society: Questioning Connections in the Digital Age*. 'It's a chilling effect.'

Tencent says that in order to comply with China's legal and regulatory requirements, it can process and handle private information without requiring user consent in certain situations.

All of this has happened in tandem with Xi Jinping's ascendance and his tightening rein on the online sphere. In 2015, the government began openly dispatching public security officers to China's largest internet companies, including Tencent. Now they had a direct line to product managers' desks and could order staff to delete content and monitor accounts much more effectively, even down to the specific words.

Two years after that initiative, Lotus Ruan, a research fellow at Citizen Lab at the University of Toronto, concluded in her research that WeChat blocked keywords related to the nineteenth Party Congress for about a year before the important political conclave. The censored topics included criticism and general speculation around the event. It also encompassed leaders, central government policies such as the Belt and Road Initiative, core ideological concepts, and power struggles within the Party.

WeChat is now much more sophisticated when it comes to censorship and minimising disruptions of user experience. Tencent has more than 20 million public accounts. It would be impossible for the company to rely simply on manual labour to police all that content. But by employing artificial intelligence, the company can automatically detect banned words, images and even screenvoice messages. For users outside of China, if their conversations entail sensitive information banned within the country, they have no issue sending content to each other. But WeChat is then able to filter out the messages and ensure they never make it across the

Great Firewall and into the country. Citizen Lab discovered that WeChat is also systematically monitoring content sent by international users, to build algorithms to better surveil people in China. The app does so by screening images and documents shared by overseas accounts, then adding the digital signature of any files deemed sensitive to a blacklist. Those files subsequently cannot be sent or received by China-registered users.

One of my friends, who had been posting information about the 2019 pro-democracy protests in Hong Kong for months, once complained to me that she had suddenly become unpopular within our chat group, because no one responded to her messages. Turns out she was the only person who could see the content she sent. Her texts and social media posts became invisible to her friends. 'I never realised I was getting censored.'

WeChat's fate – that of becoming a government monitoring tool – was set perhaps from day one. It was part of the reason why, despite the company pouring millions into overseas expansion – including deploying soccer star Lionel Messi – the app never gained traction beyond the Chinese community. Tencent chief strategy officer James Mitchell officially declared a halt to WeChat's expansion in Western markets even as the product soared past 468 million users in 2014. That was the price WeChat had to pay for being a Chinese product.

THE APP FOR EVERYTHING: MINI PROGRAMS

Because of censorship, WeChat's global ambitions were severely curtailed. That meant, for the business to expand, it had to increase its functionality and become the app for everything in China. If it could be the go-to service for 1.4 billion users in the country, that fulfilled Pony's vision of connecting people, content and physical goods and services. WeChat would still have scope to grow.

Going global is 'essentially the challenge of every Chinese company', Martin said. 'At the time we faced an important decision: whether we want to build WeChat into a platform. As soon as we focus on China, we can actually go deep in China, we can add different functionalities. From instant messaging we add Moments, from Moments we add official accounts, from official accounts we added WeChat Pay. That entire strategy going deep.

'I think there was a conscious choice. China is our hometown, we absolutely need to tackle this market first. As a result, we build it into a platform; as a result it becomes much less transferable,' he said.

Going deep in China is exactly what WeChat did. In September 2016, it sent out invites to 200 programmers asking them to design services for 'xiao cheng xu', now known as Mini Programs.

The idea was to allow users to access a raft of other services without having to actually download a bunch of apps. Tencent's gambit, which would eventually exponentially increase time spent on WeChat and link users to real-world services, kicked off the following January.

It drew scepticism at the time. China's largest search engine, Baidu, had experimented with the same concept only two years previously, only to see that product flop, in part because the company was then still finding its footing in the mobile age.

When I sat down with the Tencent director in charge of the initiative at the time, I asked whether WeChat would be creating an App store of its own to recommend and rank the lite-apps. To my surprise, the director told me they had no such plans. 'We don't want to tell people what's good and bad, we also don't want to create rankings to distort app developer incentives,' he said.

To me, that design meant Mini Programs would be a slow burn. People had got so used to the idea of App Store rankings and recommendations, it seemed hard to imagine anyone could discover what services were available within WeChat without a search bar that provided a dropdown menu for the most popular apps.

The pitfalls of rankings were obvious though. Chinese developers and merchants are masters of generating bot traffic and gaming the system to create fake activity and boost ratings. That was the exact scenario that WeChat was trying to avoid.

In one developers' seminar, Allen unfurled his vision for Mini Programs: users should ditch the lite-apps the moment they're done using them. It was controversial. Startups and developers asked why they would develop for a platform that eschewed the principle of user retention.

So Mini Programs indeed got off to a slow start. By February, Chinese media started branding it the 'Waterloo of Tencent'. An internet consultant's research showed that same month that nearly 70 per cent of developers were ditching the platform.

So Allen made a compromise, a design that would allow search functions to a certain extent but still limit the bots. WeChat made Mini Programs searchable within the app, provided people typed in the full name. Then within a week, it introduced keyword searches for categories like bikes, movies and music, making it easier for people to find the kinds of lite-apps they wanted. More importantly, by the end of February 2017, users could scan QR codes and be directed straight to the Mini Programs of services like Mobike, then a popular two-wheeler sharing service.

Mini Programs has now become so successful that its biggest competitor in the league of super-apps – Jack Ma's Alipay – announced in early 2020 that it too would invest heavily in its own system of mini programs, elevating it to one of the most important strategies for the company. Alipay – which also boasts one billion users – saw the threat that WeChat's new platform posed, and started copying three years ago the same design, though with little traction at the start.

In the meantime, WeChat was chipping away at Alipay's market share. Alipay once accounted for the vast majority of online mobile payments in China thanks to its role as the checkout counter of choice across Alibaba's online malls. But by the third quarter of 2020, its market share was down to 54 per cent. By undercutting Alipay, Tencent was effectively disrupting Alibaba itself, gaining access to its arch-foe's customers and their shopping data.

The fast-paced evolution of Mini Programs underscores a key to WeChat's success: upgrade incrementally and move fast. One WeChat employee told me that his team pushes out small changes almost every three days, and at its peak released version upgrades as frequently as every two weeks.

Momentum really picked up with the introduction of the QR code scanning function. Like most users, I found it most useful to sit down in a diner, scan a code on the table, and start ordering food directly from the restaurant's Mini App on WeChat. After the meal, I would pay on the same interface using my WeChat digital wallet. The same applied to booking movie tickets, streaming music, hailing a taxi, unlocking a bike parked on the side of the road. All of a sudden, it didn't seem so necessary – clunky, even – to use a specific app for every activity in life.

It also meant exposing a tremendous amount of personal information – far greater than Facebook or Instagram could garner. It's remarkable that a billion people have opted to sacrifice all aspects of their privacy for convenience.

In the US, the public raises concerns about Facebook's access to sensitive personal data. WeChat trumps Facebook by a long shot, in terms of both depth and comprehensiveness. As a reluctant yet avid user myself, I experience first-hand, on a daily basis, how WeChat's design, revolving around my social connections, keeps pulling me back to the app.

A question that many have posed, often triggering scepticism or uneasiness, is whether WeChat, without the burden of censorship requirements imposed by the Chinese government, could have shown Silicon Valley titans like Facebook the way to properly monetise a business with a billion users.

WALL AND EGG

Brushing aside dealings with the government, Allen can be extremely idealistic in his product design philosophy. 'What is WeChat's dream? Like I said, from an individual's perspective, to become the best tool friend for a person,' Allen said during a speech.

Internally, Tencent staff mock Facebook's business model as being primitive. 'When you have a billion users, the last thing you should be thinking about is how do I sell them a ton of ads,' one Tencent employee told me in our discussion. 'What you should be thinking of instead is how to ensure you keep these users happy, so you don't drive them away because you're annoying the hell out of them.'

WeChat has kept its dominance in a sea of rival Chinese apps in part because it has maintained a clean interface and hasn't bombarded people with ads and alerts. Take Allen's insistence on never changing WeChat's welcome screen, which shows a silhouette of a lone person looking up at a vast depiction of planet Earth. Instead of using the space to sell banner ads, Allen has been militant about keeping that page sacrosanct.

'Every time you see the loading page of WeChat, you'll think, "What is this person doing? What is he doing in front of planet Earth?"' Allen said in a speech. 'One billion users will have one billion interpretations and find something that moves them.'

He once used the metaphor of how, if WeChat were caught between a wall and an egg when it came to commercial interests, WeChat would always stand beside the egg. 'If you become big, WeChat will curtail you; if you just started, WeChat will help you,' he said.

While Allen has never talked about his or WeChat's involvement in censorship and surveillance, he takes particular pride in his team's restraint in monetising the product. In January 2021, Allen partially addressed user concerns about privacy by vowing that WeChat staff were forbidden from looking at chat records without permission. Those in violation would be fired from the company.

'User chat records are only kept temporarily on the cloud for three days; in fact, we don't store them. If we chose to conduct analysis (on user conversations) it could bring advertisement revenue for the company, but we haven't done any,' Allen said during a WeChat conference.

Allen adheres to a simple set of rules when it comes to design:

1. the product needs to be creative, it needs to have innovation;
2. it needs to be useful;
3. a good product is beautiful;
4. a good product is easy to use;

5. a good product is reserved, not flamboyant;
6. a good product is honest;
7. a product needs to be long-lasting and withstand the test of time;
8. a good product doesn't leave out any details;
9. it needs to be resource saving, not wasteful of any resources;
10. minimise design, less is more.

'For me, I am very lucky, very blessed. As a product manager, to be able to create a product that has a billion users gives me a great feeling of accomplishment. But I think what makes me even more lucky is that I can showcase my understanding of the world in the product, using it to demonstrate my values. This is precious,' Allen said publically.

His drive for simplicity arises from an obsessive personality. The engineer once managed to win the Alfred Dunhill Links Championship in golf, drawing astonishment across social media in China. 'I think there are so many obstacles in golf, I can't count them. People challenge themselves on their weakness; this is a great source of joy,' Allen said.[3]

He's surprised people repeatedly. During WeChat's 'public day' in January 2018, the company showcased a simple game that became an instant hit. WeChat was trying to make a move into gaming for Mini Programs at the time, leveraging Tencent's traditional expertise in mobile entertainment to give its newest innovation a boost. It had just released a title called Tiao Yi Tiao, or Jump Jump. The game became an overnight sensation, hooking 100 million daily active users (myself included). Players hopped from block to block, trying to not fall off and to land in the centre for extra points by tapping their smartphone screen. They needed to press longer when the gap between blocks was wider. The most addictive element for me was how my score was shared with all my WeChat contacts on a scoreboard. I spent hours a day trying to beat my friends, but managed to score only 306 points at best. It was better than average, just enough for me to take a screenshot to claim bragging rights.

---

[3] Tencent Sports.

As thousands of people sat in the exhibition hall in Guangzhou, they focused their gaze on the big screen showing the game in motion. At first, they thought it was just a video recording, but then as the player's score ticked past 400, people started to focus and root for the gamer. By the time the tally soared past 800, the audience was gasping, erupting in applause when the game finished at 967.

Then Allen stepped out from backstage.

'This isn't my best game, I once scored 6,000,' he said coyly, adding that the whole shebang was to prove to his friends that he wasn't using a software program to cheat.

His eccentric personality is the reason he's been branded one of the loneliest executives in Chinese tech, but also won him a cult-like following within the community. WeChat's Mini Programs shift enabled it to spread its wings and embed itself with offline, physical businesses such as restaurants, convenience stores and hotels, all of which are now using the service to manage membership and even their own inventory and business.

REVENUE PRESSURE

Internally, though, Allen faces mounting pressure. For all of WeChat's influence, the app itself isn't a revenue driver, and for many years it had to be heavily subsidised. Tencent's social media ads only accounted for 10 per cent of revenue as of 2014, when WeChat hit 400 million users. More than half of the company's money came from games, mostly from desktop then.

Staff at Tencent told me that senior executives have pushed WeChat to generate more revenue through ads. But WeChat product managers, proud and almost idealistic, remained resistant to the idea. The staff from the gaming division often felt they were carrying all the weight, while the WeChat unit basked in glory despite bringing in little money.

Allen has consistently pushed back against introducing ads at all, even though he eventually relented in 2015, starting with one ad per day for users. Today I still only see two ads a day, tops.

At perhaps WeChat's most critical juncture, Pony stuck with his instincts, siding with Allen's argument that online ads were not the only solution. They envisioned that WeChat could become so much more than just an ad dispenser; instead it could be the ultimate con-nector – food delivery, taxi hailing, on-demand goods like over-the-counter meds and groceries – connecting individuals with people and services.

That concept – bridging the online and offline worlds – has also been Tencent's strategy for the past decade, and conceivably for the future.

Despite Tencent generating most of its revenue from games in the past, the company has been shifting and expanding its sources of income, including cloud and enterprise services like providing membership and HR management tools for businesses on WeChat.

The app itself has become so much more than a WhatsApp-a-like, steadily morphing into an immense online bazaar that, by the end of 2020, was funnelling $240-billion-worth of business in goods, games and services like ride-hailing and food delivery. It would become Tencent's impossibly large window to China's consumers, a giant platform through which it can sell, market, draw and connect users and ignite online inter-actions. In short, it's become synonymous with Tencent's steady ascent into the upper echelons of the global internet industry.

In the process of building WeChat into a connector, Pony had to also ensure that there were enough reasons people would open the app in the first place. Ordering food on Meituan was one of them. But the other function that both Pony and Jack Ma foresaw as essential for their companies' success was ride-hailing. Calling for taxis, high-end cars and pooling with other commuters is deemed as a high-frequency business – on average a person needs to use the service at least two times a day – a perfect way to keep users engaged.

Both Alibaba and Tencent saw the potential in ride-hailing and how it could benefit their mobile payment services. And both billionaires went after the same strategy, again backing a third-party champion they thought could promote their digital wallet ecosystems.

The subsequent war would end up with billions burnt, China's two largest ride-hailing giants – emerging victorious from among hundreds of contenders – merging into a single entity, and eventually kicking Uber out of China.

CHAPTER 8

THE RIDE-HAILING BATTLE

In 2013, Tencent's head of investment, Richard Peng, tracked down the founder of a scrappy ride-hailing company called Didi. He had spent months trying to set up a meeting, and when the round-faced, soft-spoken founder Cheng Wei seemed hesitant, he invited the young CEO into his office. Then he locked the door.

'Buddy, first of all, I know you don't want our investment,' Peng told the founder, knowing that Cheng once worked for Tencent's nemesis, Alibaba. Then he added: 'Secondly, I have to invest.'

Ride-hailing by then had become the new 'it' thing. Before its emergence, commuting in big cities could be a real nightmare. If the traffic congestion, the torrid summers, freezing winters and pollution wasn't enough, getting a taxi was an ordeal, especially in the capital Beijing. Illegal taxis that operated without a government licence were rampant, always lurking at big transport hubs like train stations and bus terminals, as desperate passengers dismissed the dangers of getting into cars of complete strangers, without any method of tracking their itinerary.

The advent of ride-sharing apps changed everything. During one particularly harsh winter in Beijing, a big snowstorm bolstered the popularity of such services after passengers struggled to find taxis during the busy holiday season. Suddenly, hundreds of car-hailing companies mushroomed overnight, creating colourful apps that became a mainstay on peoples' smartphone screens. They became the weapons of choice in the severest cash-burning battle of the technology world.

Google Ventures put $258 million into Uber that year, its largest deal ever. But in China, Didi emerged as one of the most promising contenders.

Today, Tencent's alliance with Didi, particularly through WeChat, is one of the most important in the tech giant's portfolio of investments, apart from those within its core gaming business. Securing that deal gave the internet giant the wedge it needed to disrupt the financial payments sector and break the market dominance of billionaire Jack Ma's Alipay, which at the time controlled 90 per cent of online payments in the country.

Tencent, known for its tenacity in vanquishing competitors, battled Jack Ma using ride-hailing as a proxy before finally joining hands with the enemy. Yet it was Tencent that eventually came out the bigger winner. To tell the story of this epic fight, we need to trace the journey of the young entrepreneur, Cheng Wei.

'It's a long story,' Cheng said, sitting in his office with views of the western hills in Beijing. 'It had unexpected twists.'

FOOT MASSAGE PARLOURS

Cheng was born in May 1983 in Jiangxi, a landlocked region in eastern China famous for being a cradle of Mao Zedong's communist revolution. When I first met the entrepreneur, he was already the cherubic and bespectacled face of the nation's ride-hailing leader, a visage that wouldn't look out of place in a video game parlour at 2 a.m.

His father was a civil servant, mother a mathematics teacher. He says he excelled at numeracy in high school but botched the most important test of his life – the university entrance exams, through which the country winnows two-thirds of the 10 million people that complete the test annually. Cheng forgot to fill out the last page of his maths quiz. He scored only 72 out of 150 points for his best subject. His mother, an associate professor, proved more than supportive.

'She was shocked for a second, but she immediately said, "Don't worry. Let's go back home and eat. You need to take the chemistry test in the afternoon,"' Cheng recalls. He ended up doing pretty well in the rest of the subjects, scoring 50 points higher than the first-tier university benchmark and winning a spot at a college in Beijing, still a remarkable achievement for a student in impoverished Jiangxi.

Instead of staying in high school another year and retaking the exam, Cheng opted for the less prestigious Beijing University of Chemical Technology. He signed up to study information technology, but was reassigned to business management. 'Who would let a young graduate do management? So I tried a lot of jobs,' said Cheng after he graduated.

He tried working as a tour guide. He went into life insurance. He clung to that job just half a year and didn't manage to sell a single policy. He even sought help from his teacher. 'I asked my teacher: "Are you supportive of me doing such a challenging job?" And my teacher said, "Of course."' But when Cheng tried to convince him to buy a policy, his teacher responded that even his dog was insured.

At a job fair, he applied for an opening as a manager's assistant at a company billing itself as a 'famous Chinese health-care company.' But when he showed up for work in Shanghai, luggage in hand, he discovered it was actually a chain of foot massage parlours.

'That's why Didi seldom does advertising, because I think it's all a scam.' He stayed at the national chain for a few months. 'I stood there with my luggage for about one minute thinking, should I take my bags and go back to Beijing and tell people I got scammed. Or should I go up and try it.'

Till this day, he's not a fan of foot massages.

IRON ARMY

It was around this time that Cheng began to think hard about his future. And one day the answer struck him: the internet was his way out.

He showed up uninvited at Alibaba's offices in Shanghai and told people at the front desk he wanted in. He landed a job in sales, earning 1,500 yuan or $225 a month. 'I am very thankful towards Alibaba,' Cheng says. 'Because someone stepped forward, didn't shoo me away, and said, "Young people like you are what we want."'

Despite his earlier insurance fiasco, Cheng excelled at selling online ads to merchants. For the next eight years, he worked for Alibaba's toughest team, the sales cohort known as the 'Zhong Gong Tie Jun', meaning central supply iron army – a pun pronounced the same way as the Communist Party's Iron Army. The team was known for its aggressive approach to enlist merchants to sell on the e-commerce platform. The experience laid the foundation for Didi's success, for one of the very first challenges the company faced later would be convincing reluctant taxi drivers to sign up. Cheng reported to an outspoken and exuberant Alibaba veteran named Wang Gang. When he first met Cheng, says Wang, his sales numbers were strong, 'but his real strength was being an MC' at company events for customers.

In 2011, Wang and a group of underlings, including Cheng, started to brainstorm ideas for startups. Wang says he had grown disgruntled with Alibaba after getting passed over for a job running a unit at Alipay, the company's fast-growing payment division. The group came up with ideas for starting companies in education, restaurant reviews, even interior decorating.

HONK HONK TAXI

In early 2012, Wang and Cheng saw a smartphone app called Momo, a Tinder-like dating app that allowed people to identify the location of other users on an online map. Wang says that the notion of tracking attractive females on phones piqued their interest in the enormous potential of a smartphone's GPS capabilities.

The duo's attention turned to a foreign ride-hailing startup that was raising money at a rapid clip and planning to spread around the world. It wasn't Uber; its name was Hailo – a UK-based company that worked with London's iconic black cabs, driven by licenced drivers. Cheng Wei broke off from the group to study Hailo and judged that its model could be replicated among China's two million taxi cabs. Wang became his primary financier, investing 800,000 yuan (about $119,000) of the 40 million yuan he had made at Alibaba.

Sitting in a café across from an Apple store and looking for a name just as memorable, Cheng called his new company Beijing Xiaoju Technology Co.: 'xiaoju' means little orange. He dubbed its taxi-hailing application Didi Dache, or 'Honk Honk Taxi'.

Cheng and several colleagues initially set up shop in a shabby 100-square-metre warehouse space with a single conference room in the northern part of the city, near Beijing's fourth ring road. It was close to the top universities, making it easier for them to poach talent. The team called each other classmates, just like they did at Alibaba, because they worked tirelessly and thought creatively about solutions, and within the company Cheng was referred to as laoda or 'big boss'. They spent the first months trying to out-manoeuvre the dozens of other ride-sharing companies that had latched onto the same idea.

'By the time we were actually about to launch the service, about thirty companies emerged at the same time,' he says. 'There were different models. Some companies were much more powerful than us.'

Cheng dispatched two of his first ten employees to begin operations in Shenzhen, home to Tencent and Foxconn's iPhone factories, because he thought the city had the most liberal regulatory regime in China. Didi was promptly suspended by local authorities for regulatory violations and told to rectify its operations before it could resume business.

'In the beginning, we needed a lot of courage,' said Cheng, adding that this was the reason why in the early days the founding team's favourite song was 'The Brightest Star in The Night Sky'. 'This was a long journey, there would be journeys in the dark, but the brightest star would guide you. You can't see a business model, you can't see yourself being legalised, you can't see the end of competition.'

But Didi, it turned out, had several advantages over many of its competitors. Some copied Uber's initial strategy in the US and tried to enlist China's limousine and town-car chauffeurs. There were far fewer high-end cars than yellow cabs and it wasn't an effective approach. Cheng and his colleagues applied the spirited camaraderie and ferocious work habits they developed at Alibaba to the problem.

Cheng placed a great deal of emphasis on funding from the start. But that avenue seemed closed. Even before the company had a product, he reached out to investors – only to be rejected by more than twenty firms. 'The biggest lesson for me was that everyone told me, "You're great." And then there would be no follow-up.'

The budding company switched tactics. When YaoYao, a rival backed by Silicon Valley's Sequoia Capital, won an exclusive contract to recruit drivers at Beijing's airport, Didi employees descended on the city's biggest railway station to promote their app. Because the fledgling company had but a meagre $119,000 in the bank at the time, its small founding team had to do everything themselves, including personally helping drivers install the app. The team hunkered down at train stations to chat up drivers as they waited in line for their next customers.

'Drivers don't have one or two hours to wait. They only have about fifteen minutes. They're moving along the road as they wait in line. It's relatively dangerous and difficult,' said Cheng. But that winter, Didi was able to bring onboard about ten thousand drivers, which meant most of the cabbies who had a smartphone in Beijing.

'The most important thing is to drive yourself to a state of despair, and then God will open a window for you. So up till today, we think that money is just one element of resources, but it is never the most important. There will always be someone who is richer than you. The most important thing is your persistence,' said Cheng.

Instead of giving away smartphones to drivers, an expensive proposition for a capital-strapped startup, they focused on providing their free app to younger drivers who already had phones and were likely to spread the word about Didi. Then their opportunity came.

SNOWSTORM BLESSING

During an epic Beijing snowstorm late in 2012, when it was impossible to hail a cab on the side of the road, residents turned to the app and the company surpassed a thousand orders in a day for the first time. 'If it didn't snow that year, maybe Didi wouldn't be here today,' Cheng recounted.

The blizzard helped in another way as well. Didi was making no money – riders were hailing cars over the app but still paying yellow-cab drivers in cash, so there was no way for the company to take a commission. The young entrepreneur was certain he could develop a revenue stream after he built a customer base – Alibaba had followed a similar blueprint, after all – but the rejections from investors piled up.

Then Allen Zhu, a partner at Beijing-based GSR Ventures, heard about the startup after the blizzard and cold-called Cheng after reaching out on Weibo, a Twitter-like service. Zhu recalled a nervous and stressed-out Cheng continuously blushing during those early fundraising meetings. But GSR put in $2 million, valuing Didi at a paltry $10 million.

'After the first snowstorm, we found our first investor. That's when I knew investors will not provide charcoal in snowy weather, they will only add highlights to things that are successful,' said Cheng, invoking an old Chinese idiom that in modern-day startup lingo means investors won't help you when you're down, only when you're doing great.

Then, some bad news: Alibaba invested in a rival ride-hailing company, Kuaidi Dache ('Fast Taxi'). Cheng was only too aware of how the success of startups in China depended on the strength of connections to either of the Big Two – Alibaba and Tencent. That's when Cheng reached out to Richard Peng again.

PENURIOUS ORIGINS

Richard Peng, who was Tencent's head of merger and acquisitions, knew he already wanted to invest in Didi, but still wanted to see what the founder was made of. So during his meeting with Cheng, he stress-tested Cheng with two important questions. First, he challenged Cheng's assumptions about the significance of a car-hailing app. Cheng responded by saying, 'Richard, you're wrong,' and told him that increasing efficiency in getting taxis and rides in others' cars was the future. He posed a maths calculation: if China's car-hailing market was worth 600 billion yuan ($85 billion) a year, since every ride costs 20 yuan and there are 30 billion orders annually, pricing a ride at 1 yuan equates to a 30-billion yuan market.

'Cheng was talking about the business from a global perspective. He was logical, methodical and also a big-picture thinker. He also wasn't hesitant in pointing out I was wrong; that impressed me,' said Peng.

He also took a liking to Cheng's personal story of overcoming hardship, something close to his own heart. He was one of nine children born to farmers in the same province that Cheng grew up in. Peng still speaks with a trace of a lilting Jiangxi accent, hinting at a penurious rural origin. Giving up on vocational school, which would have meant starting to make money earlier, Peng chose instead to go to high school. As neighbours looked on during one particularly tense family meeting, Peng convinced his father: 'All my life you've told me to be someone and do something important with my life. This is what I want to do now.'

That set Peng on a path towards what seemed at the time like a routine academic and business career. He got into Beijing's Tsinghua University – alma mater to Xi Jinping and countless other party cadres – where he sought inspiration in the writings of Alvin Toffler and business icons Armand Hammer and Lee Iacocca. He became something of an entrepreneur: early penny-ante undertakings included bicycle repair and finding work for construction workers.

One of his earliest jobs was at the Shanghai stock exchange, a plum posting he forsook in 2001 to attend the University of Pennsylvania's Wharton School. After getting his MBA, Peng joined Samsung Group before leaving in 2005 to become Google's first local Chinese hire. That was when he caught the attention of Tencent President Martin Lau. He parted ways with the search giant in 2008 to join Tencent – a company then one-sixteenth the US company's size.

As head of investment and acquisitions at Tencent, Peng proved a prolific deal-maker, often forcing companies to sign term sheets the same day it extended an offer, to avoid price manipulation.

Peng's second challenge for Cheng was asking the entrepreneur, what was the key to making his business successful? Most of the other CEOs he had spoken with – Didi was fighting some two hundred outfits at the peak of the competition – would talk about how to get more drivers to join or expand marketing efforts. Cheng, who had never written a line of code in his life, surprised him by saying in 2013 that the key was algorithms.

Peng sensed Cheng had found the key to making his company better than all the others. He assigned a $60-million valuation to Didi, 50 per cent more than what Didi was seeking, to ensure no one else would snatch up the company.

SEVEN DAYS, SEVEN NIGHTS

With the backing of two rival Chinese internet giants, Didi and Kuaidi trained their sights on each other. During one notoriously difficult week, reverently known at Didi as 'Seven Days, Seven Nights', both companies battled intermittent technical problems, sending drivers and riders scurrying from one service to the other and back again.

Engineers were holed up in Didi's cramped offices for so long and worked so hard to resolve their issues that a classmate didn't visit his pregnant wife until she went into labour. Another had to have his contact lenses surgically removed.

'We were mad, but because we were so pumped up, it masked the fact that we were all so exhausted,' said Li Hairu, an early employee.

Thanks to that meeting with Richard Peng in 2013, Cheng had an ace up his sleeve. He called Tencent and Pony for help. The billionaire agreed to lend him fifty engineers and a thousand servers, and invited Didi's team to work temporarily out of Tencent's more comfortable offices.

But that just proved a temporary reprieve. Didi still wasn't making any money, and Cheng needed to raise more capital. Local investors were reluctant to chip in more. He had to head to the US. 'We had burnt a lot of money,' he says. 'Investors were like, "Whoa."'

Cheng arrived in New York on Thanksgiving in 2013. It wasn't cold in Beijing at the time, so he wore a shirt for his trip. Turns out it was snowing in America. The meetings were rushed and poorly prepared. He met with four investors, and failed to impress any. One agreed, only to bail out.

That night Cheng was supposed to fly to San Francisco. He called an Uber in New York but missed four planes, because of the snow. By the time he finally arrived in San Francisco, it was after midnight. The second day wasn't successful either. 'I was very depressed after I came back to China.'

RED PACKETS

In early 2014, during the Lunar New Year, everything changed. Tencent ran the successful so-called Red Packet promotion over the holiday, which allowed WeChat users to send small financial gifts to friends and families using their smartphones. It was a smash success and led Tencent to an epiphany: mobile payments were the future.

Now Tencent was even more determined to throw all its weight behind Didi. A lot of thought went into the design behind linking WeChat and Didi to encourage driver adoption. WeChat placed Didi on its services landing page, channelling its enormous army of users to the ride-hailing app. In return, Didi adopted WeChat's then-nascent payments system, in an attempt to transform drivers' age-old habits of only accepting cash. 'It looks intuitive now, but at the time it was extremely hard,' said Li, the early employee.

Didi introduced a design feature of its own to help WeChat Pay avoid the fate of earlier failed mobile wallets. Traditionally, it would take twenty-four hours before drivers could receive the money they earned in their bank accounts. Didi and Tencent shortened that lag to under two hours, paying drivers upfront with their own money and building its own clearing network with banks.

Tencent also backed Didi with capital when it subsidised users to boost usage. They gave people 10 yuan ($1.4) per ride. 'People soon began saying, "How could you not use WeChat payment while hailing a cab?"' Cheng exulted.

Alibaba responded in kind, pumping money into Kuaidi, which was integrated with its own mobile payment service, Alipay. Together, the companies spent about 2 billion yuan on discounts and subsidies to customers on the twin taxi-hailing apps over the first few months of 2014. Ridership soared.

Li Hairu recalls checking daily orders every midnight during those months – 3 million! 3.1 million! – and yelling them across the office to her colleagues.

TWO HOURS' SLEEP

Advanced cloud computing was non-existent back then. Everything had to be manually tallied via Excel spreadsheets. Li, who often worked in the office till 2 a.m., would go home to work on WeChat and emails until 5 a.m., then head out again after just two hours' sleep.

Working as part of a small startup, with about a hundred people then, Didi's early hires didn't earn much and lived far from the office on the outskirts of Beijing. To save money, Li would take public transit daily rather than cab it, ironically.

The stress also put marriages at risk, including Li's own. When her son was in third grade, the school required parental pickups. Usually that responsibility fell on the nanny, but on a few occasions, when it was Li's turn, she would forget. Her son would wait for hours in an unheated waiting room at his school during the depths of a freezing Beijing winter. When the boy called, she'd agree to notify her husband, hang up the phone – then promptly forget to do so. After this happened multiple times, 'my husband got so mad at me,' said Li. 'We were facing explosive growth at the time, so we had to just suck it up and stick it out, otherwise all our efforts would have gone to waste.'

In 2014, Cheng made one of his most important hires. Jean Liu, the daughter of Liu Chuanzhi, a first-generation Chinese tech mogul and the founder of PC maker Lenovo. Jean had worked at Goldman Sachs and became a managing director in principal investment after studying computer science at Harvard.

Jean had in fact tried previously to invest in Cheng's endeavour. 'I wanted to invest in Didi a year earlier but it didn't work out,' she told me during our first meeting in Hong Kong. Already a mother of three, Jean wasn't afraid to flaunt her femininity, wearing a crimson dress, peep-toe pumps and side-swept bangs.

She tried again in 2014 and that again fell through. When she met Cheng on 19 June that year, he told her, 'You're too late.' So she said: 'Well, since you won't let me invest, how about you let me work for you?' And Cheng Wei said, 'Sure.'

TIBET TRIP

They set up a lunch meeting on 22 June and within three days, she'd agreed to come on board. The first thing she did was join nine other Didi members on a ride to Tibet. 'Sitting with those people in the vastness of the inky-black void, we talked about life and ideals. If you were going to live, how would you live? We were all pretty idealistic, full of passion,' she said.

The trip itself was in reality more precarious. Being severely underprepared, the group of ten mistimed their trip, driving too far beyond the last village and ended up on switch-back roads in the mountains. In their hunt for a lodge and a bathroom, Cheng and Jean also ended up getting chased by mutts bred by local farmers. But the expedition served its purpose.

'We bonded in the mountains,' said Jean, who got up every day at seven, jogged and did yoga for forty-five minutes, after often only sleeping four hours a day when she first joined.

Jean and Cheng often worked side by side until midnight. Cheng would sometimes lighten the mood by saying: 'Hey, how about we listen to "little apple"?', a 2015 Chinese song by the Chopstick Brothers that swept the country much as Baby Shark would years later. Cheng could at times also be playful, bragging in the office how he could hold a full lotus position with both of his ankles on top of his knees.

Part of the reason Jean was so determined to join Didi stemmed from her own frustration with China's transport system. One time, she and Cheng went to meet a transportation bureau official. 'It was one of those situations where you absolutely could not be late,' she said. When the duo got stuck in Beijing traffic, they jumped out of their car, walked for a mile, dashed into the subway and then hailed a tuk-tuk, a makeshift vehicle illegal in many parts of the city.

'I was coming down with a fever that day. After all that hassle, my makeup was melting. It felt as if I'd come out of a boxing match, a gym workout,' Jean said. 'You're like a sucker when you're travelling in China in traffic. No matter where you go, you don't feel dignified.'

ENTREPRENEURIAL GLADIATORS

What Jean brought to Didi was a global perspective. She introduced a rotation scheme to help general managers understand how each team worked. More importantly, she wasted no time in focusing on her most important task: fundraising.

In the words of Dexter Lv, founder of Didi's rival, Kuaidi: 'I was so surprised Jean was able to fundraise so efficiently. She practically scooped up all the money there was to be had on the street.'

Indeed, the funding battle had become so intense that investors – many of them global investment houses with the backing of US pensions and endowments, including Tiger Global Management and Russian billionaire Yuri Milner's DST – became concerned that the cash-burning war wasn't sustainable, especially given word on the street that Uber was sizing up its own Chinese entry.

The investment model of burning capital, doling out subsidies to win scale and build a defensive moat, was no longer making sense from a return-on-equity perspective because the competition prevented them from monetising their business. In order to maintain the users, one of the two camps would inevitably have to make the first move and ask users for more money.

Or merge. Both sides eventually turned to veteran banker Bao Fan to broker a deal.

As with the Meituan–Dianping deal he brokered, Bao took fees from both Didi and Kuaidi as sole arranger.

The key, Bao said, was overcoming egos. 'These guys were entrepreneurial gladiators who never shied away from a fight,' he said, reclining in a high-backed leather armchair in the Beijing headquarters of his boutique investment bank. 'But they were killing each other. I wanted them to make peace, not war.'

Bao resorted to stealth. He code named his merger-prodding strategy 'Project Rush,' after the 2013 film in which Formula One racing rivals James Hunt and Niki Lauda ultimately forge a friendship based on mutual respect. Lv's company, Kuaidi, was Lauda. Cheng's Didi stood in for Hunt. Their powerful shareholders, internet giants Alibaba and Tencent, were Ferrari and McLaren, respectively. 'It was super-confidential,' says Bao.

Bao embarked on his mission, by persuading the founders they were reducing their own equity piecemeal by tapping cash from outsiders to sustain their rivalry. They listened. On 14 February 2015 – Valentine's Day – the companies agreed to matchmaker Bao's plan to wed in China's biggest internet merger.

Didi, which had more volume than Kuaidi, ended up controlling 60 per cent of the combined company, which Cheng insisted on running as a condition of the deal. Didi agreed to incorporate Alibaba-backed Alipay on its app, but the damage from WeChat had already been done.

Before Didi incorporated WeChat Pay, Alipay accounted for more than 90 per cent of the online payments market. Thanks to Didi, WeChat Pay made a breakthrough in mobile payments and – just like that – its market share was then closing in on half.

UBER INVASION

Cheng had another reason to agree to the merger. Word had spread that Uber was ready to roll out in China, and the local players needed to corral all the ammunition they had to prepare for the bigger fight.

In late 2013, Travis Kalanick and a team of Uber executives toured China to size up prospective partners and rivals. They visited Didi's offices. Cheng kicked things off by telling Kalanick, 'You are my inspiration,' after which the mood became tense. Emil Michael, Uber's then senior vice president for business, remembers what may have been some psychological warfare: 'They served us maybe the worst lunch I've ever eaten,' he says. 'We were all just poking at our food, wondering, is this some kind of competitive tactic?' (It wasn't. Jean Liu, Didi's president, later apologised to Michael for the food.)

At one point during the meeting, Cheng walked over to a whiteboard and drew two lines. Uber's line started in 2010 and went up sharply and to the right, depicting its rapidly rising ride volume. Didi's started two years later, in 2012, but had a steeper curve and intersected Uber's line. Cheng said Didi would one day overtake Uber, because China's market was so much larger and many of its cities restrict the use and ownership of private cars as a way to manage traffic and pollution.

'Travis just smiled,' Cheng recalls. Kalanick raised the possibility of Uber investing in Didi, but he demanded a 40 per cent stake, says Cheng. 'Why would I take it?' he replies, when asked if he considered the offer. Uber's executives were impressed. Kalanick, Michael says, 'told me that among all the ride-sharing founders, Cheng Wei was special. He was just a massive cut above anyone else in the industry.' Uber and Didi were going to have to settle the matter in the market.

By the beginning of 2015, it seemed that Uber had an insurmountable advantage. It had a better app, powered by more stable technology. Investors valued it at $42 billion, about ten times Didi's valuation at the time. While Didi was consumed with its merger with Kuaidi, Uber was catching up: it controlled almost a third of the private car-hailing market in China within a few months. 'At that time, we felt like the People's

Liberation Army, with basic rifles, and we were bombed by airplanes and missiles,' Cheng says. 'They had some really advanced weapons.'

WOLF TOTEM

Cheng is a student of military history, particularly such heroic events as the battle of Song Shan during the Second World War, when Chinese nationalist troops tunnelled under a mountain to surround the invading Japanese army. Cheng held morning meetings with his senior staff, which he called the Wolf Totem. The name, based on a popular novel set during the Cultural Revolution about urban students sent to live in Inner Mongolia, connotes aggression. The Wolf Totem meetings studied Didi's daily results and adjusted the amount of subsidies given to drivers and riders. Cheng would regularly warn employees, 'If we fail, we will die.'

In May 2015, Cheng went on the offensive. Didi said it would give away 1 billion yuan in rides. Uber matched it.[1] Cheng and his advisers searched for ways to fight the American company on its home turf. Uber, they reasoned, was like an octopus – its tentacles were everywhere in the world, but its mantle was in the US. Wang, the early investor and former board member, suggested at a meeting that Didi 'stab Uber right in its belly'.

Wang says Didi contemplated expanding into the US. Instead, in September 2015, it invested $100 million in Uber's American rival, Lyft. According to Wang, it was less about undermining Uber than about gaining negotiating leverage. 'The purpose of them grabbing a lock of our hair and us grabbing their beard isn't really to kill the other person,' he says. 'Everyone is just trying to win a right to negotiate in the future.'

It was widely suggested in press accounts that the Chinese government helped Didi in its battle against Uber. Cheng rejects that, noting that as the largest ride-hailing company, Didi had to shoulder most of the

---

[1] The subsidies gave rise to a cottage industry that helped drivers use modified smartphones and software to place fake bookings and trick the ride-hailing companies into paying out cash for phantom trips.

regulatory burden and paid tens of millions of yuan to cover driver traffic citations and other fines. He also points out that state-connected companies such as Guangzhou Automobile Industry Group and China Life invested in either Uber directly or its China operations.

At the peak of hostilities, Didi and Uber were each burning through more than a billion dollars a year, dishing out unprofitable subsidies to drivers and riders. Both companies were desperate for new capital. Apple invested $1 billion in Didi in May 2016. A month later, Uber raised $3.5 billion from Saudi Arabia's Public Investment Fund. The message to both sides became clear: they were going to have to wage this destructive money-losing battle for a very long time.

Cheng says the initial call for peace came from Uber; Michael from Uber contends that the Saudi money forced Didi to the table – because the investment suggested there was simply no end to the capital Uber could tap. Regardless, both sides agreed it was time to stop the bloodletting and focus on building their businesses. 'It was like an arms race,' Cheng says. 'Uber was fundraising, we were also fundraising. But in my heart I knew our money needed to be put into a more valuable field. This was why we were able to join hands with Uber in the end.'

Michael and Jean Liu hammered out the deal terms in two weeks and then met Kalanick and Cheng at a hotel bar in Beijing to raise glasses of baijiu, a traditional Chinese spirit made from sorghum. Over drinks, the CEOs spoke of mutual respect and their admiration at how hard both sides had competed. 'We are the craziest companies of our times,' Cheng says. 'But deep in our heart we are logical. We know this revolution is a technology revolution, and we are just witnessing the very beginning.'

Ultimately, the US giant agreed to withdraw from the field, selling its Chinese business to Didi in return for a significant stake in Cheng's company. Onlookers immediately credited Didi with the victory – like Alibaba and Tencent before it, Didi had officially joined the select club of Chinese corporations that had bested better-funded American rivals.

Indeed, kicking Uber out of China was just one milestone in Didi's herculean effort to survive. The foundations it laid for getting Chinese users to pay for services via mobile apps, particularly WeChat, paved the way for future innovation in fintech by the country's internet titans.

Behind Didi's big victory, Tencent stood as the ultimate beneficiary. It had won a significant stake in China's go-to app for all things ride-hailing, used Didi to buoy its market share for mobile payments on WeChat, and ultimately proved its strategy of acting as kingmaker and investor could work, encouraging many more startups to take its money and collaborate with the social media giant.

CHAPTER 9

A GAMING CRUSADE

It was the October peak of the pandemic in China. The screams of 6,000 people packed into the Shanghai stadium swelled to a crescendo. Pop megastar Lexi Liu pounded out her latest hit, joined by K/DA, a virtual pop group whose members included a nine-tailed fox mage and ninja assassin.

Scrawny and mostly bespectacled, sixteen young men walk onto a stage straight out of Tron – and the battle begins. About 45 million tuned in online, about the same amount of people that viewed the six games of the 2020 NBA Finals on TV, to watch the League of Legends world championship – a watershed moment for competitive gaming and the culmination of years of planning for its organiser, Tencent.

To understand how big gaming has become, the industry is estimated to have generated revenue of $176 billion in 2021, almost five times the box office of the movie industry. An estimated 2.9 billion people play games on some form of digital device, researcher Newzoo estimates. That's nearly one in three people on the planet.

China has been a key driver of that growth. The country overtook the US as the world's largest gaming market in 2017. That trend accelerated during Covid, as people turned to online games during lockdowns. Mobile games have become especially popular, generating more than 75 per cent of gaming revenue in China in the first half of 2020, according to the Game Committee of the Publishers Association of China.

Tencent, with key titles including Honour of Kings, and the mobile versions of Call of Duty and PlayerUnknown's Battlegrounds, is at the centre of all that action. It's also behind the studios that have created the biggest global titles, including Fortnite, League of Legends, World of Warcraft and Clash of Clans. Even today, after diversifying its business into social media, entertainment and cloud computing, the company still generates a third of its revenue from gaming.

It's been a long journey for Tencent. The gaming goliath's origins in gaming and e-sports were humble. Right around 2004, Tencent began its second attempt at gaming, after flopping at an initial foray into desktop PCs a few years previously. Despite some success in online poker and board games, Tencent didn't make a serious attempt to become a major distributor until around two years earlier, when it dispatched a small team to ferret out a potential game it could licence and distribute in China. The company locked eyes on the much-hyped Turf Battles, a multiplayer online role-playing game developed by South Korea's Imazic. Known for 3D rendering underpinned by the state-of-the-art Unreal II engine at the time, Tencent counted on the game's pre-eminence to last three years.

It didn't. The game's heightened demands on internet connections and servers wreaked havoc at Tencent. Then a whiz-kid called Mark Ren stepped in. One of Tencent's earliest employees, Mark asked that he be given a chance to lead the charge in gaming development in 2004.

Mark grew up believing that he had two destinies in life: coding and gaming. He took part in programming competitions as a teenager, and one of his proudest moments was when he won a 20-yuan award for a flight combat game in middle school. That obsession stuck with him throughout university and he became a coder at Huawei after graduation.

It was a cushy job with good benefits, but Mark had bigger aspirations. He wrote a card and board game in his spare time, hoping he could sell it to Tencent, which was just taking off at the time.

Pony showed little interest in his product but was intrigued by the young programmer with a jagged bowl cut and a prominent mole between his eyebrows, emphasising a determined gaze.

The Tencent founder invited Mark into his office and asked what it was like working for Huawei, to which the latter responded that programmers were chained to their desks at work, and it was impossible for people to brainstorm ideas.

On Mark's way out, Pony asked him if he would consider working for Tencent instead. It took one sleepless night. He agreed to join in the morning.

The youth came onboard as a team leader under Tencent's chief technology officer at the time, Tony Zhang. His passion, however, lay in coding. He requested transfers to a team on the social media platform QQ three times, but got rejected because his bosses believed he was wasted if he only focused on programming.

So when Tencent prepared a second assault on the gaming market, Mark put his hand up to lead the team. He reviewed Tencent's mistakes, concluding that the company wasn't ready to venture into games that required significant bandwidth and maintenance. He started, instead, with easy ones – like the card and board games he designed at school.

The company faced two big competitors at the time: Shanda and NetEase. Neither seemed easy targets. Shanda's most popular title had more than 60 million paying users at the time. When it listed in 2004 on the Nasdaq, its founder Chen Tianqiao became the richest person in China.

In order to win market share, Mark needed to figure out a way to leverage Tencent's existing strengths. The answer lay within QQ, which by then already had 200 million users. He managed to get QQ's team to add casual, easy-to-play card and board games to its interface, enticing users who by then had made it a habit to turn on the chat service the moment they logged online.

Learning from his competitors, Mark concluded that there were two ways to set up a gaming operation: he could either separate the development and operations teams, or he could combine them. The latter meant it would take longer to create a functioning unit. The people in charge of the teams were often from tech backgrounds, with zero operational experience. But Mark reasoned that combining forces would generate better results in the longer term.

His unit made small tweaks to designs to improve user experience. They allowed people to log on with their QQ accounts, saving them the pain of registering twice. They also applied Tencent's avatar strategy, allowing gamers to dress up in the virtual world after making purchases using so-called Q coins. Their killer function was born when they started alerting people about the games their friends were playing, instantly enhancing interaction. People could click on the notice and be directed straight into game rooms. Many years later, Tencent would recycle that magic formula when it created breakout mobile game Honour of Kings, alerting WeChat users when their friends were duking it out online.

The results surpassed Mark's wildest dreams. His team originally agreed to have a dinner celebration every time user numbers surpassed another 10,000. They did it thirteen times in total before giving up because they couldn't keep up.

Tencent's competitors were less impressed. Lianzhong's founder bemoaned that Tencent had blatantly ripped off its operation and business model. To the untrained eye, it was even hard to tell the difference between the two companies' gaming interfaces. Copycatting was rampant in China. And Tencent didn't just stop there.

In 2005, it set its sights on one of the most popular games at the time: Crazy Arcade, otherwise known as BNB. The free-to-play game was developed by Korean company Nexon and licenced to Tencent's rival NetEase. It took China by storm. At its peak, more than 700,000 users were playing online at the same time, making it one of the most popular games on the planet. I remember my cousins enthralled by a game that required players to kill their opponents by catapulting water balloons that moved horizontally and vertically. One of them spent so much time on the game, he got a beating from his parents for neglecting homework.

Tencent wanted the same, especially since Crazy Arcade's simple gameplay and low entry barrier appealed to the same user base as QQ's

– young people with lots of time to kill. Mark and his team examined Crazy Arcade, fine-tuning their own version, adding more maps to spice up the game, and injected personality into characters by letting their heads bob. But when Tencent's own game went live by the end of 2004, they realised all these add-ons interfered with the experience, making the game seem glitchy. It struggled on debut and users hovered below 10,000 for the first few months.

They went into overdrive, fine-tuning their version with multiple updates. Mark and his team never left work before 10 p.m. during that period. By July, they finally came up with a smooth and clean game, helping the company win over 100,000 users the following quarter.

Then more trouble ensued. In September 2006, Nexon sued Tencent for copyright infringement and improper competition in Beijing. The Korean company accused Tencent of ripping off thirty-seven images, game play, accessories, background colours and design – even using characters with similar names. It became the first lawsuit that involved an international corporation suing a Chinese internet company, drawing widespread attention.

The case lasted six months, but in the end the court ruled that Tencent had not violated copyright nor competed improperly. That emboldened Tencent. Even Pony said in an interview with local media at the time: 'I don't blindly innovate. Microsoft, Google are both doing what others have been doing. The smartest approach is to learn from the best examples and then try to surpass them.'

Tencent's own legal team established quite the reputation, growing into a 200-plus-strong team with experts in copyright, mergers and acquisitions. Its proclivity to win earned its legal unit the nickname 'Nanshan Indomitables', after the district they are headquartered in.

The dispute with Nexon, however, did prompt Tencent to re-evaluate its strategy and explore alternatives. It started with licensing the rights to CrossFire from Neowiz and Dungeon Fighter from Neople around the end of 2007. This was an astute move as those two games became staples for Tencent, vaulting it from a casual game operator into the big leagues during the desktop gaming era.

CROSSFIRE, DUNGEON FIGHTER

CrossFire, a first-person shooter, features two mercenary corporations named the 'Black List' and 'Global Risk', vying in an epic global conflict. Its debut was less than satisfactory when it went live in South Korea in May 2007. But due to the prior success of similar games, Neowiz demanded a licensing fee of $50 million. That sparked a huge debate among Tencent's executives, many deeming it not worth the price. Mark was adamant that the game had great potential – China had yet to experience an established first-person shooter – and he was confident that the genre could become a hit.

Mark also created a backup plan just in case CrossFire fell through. He bet on Dungeon Fighter, a multiplayer role-playing combat game that had ranked among the top ten South Korean titles for two years.

When they went live in early 2008 in China, Tencent reckoned that if it could get 300,000 people playing at the same time, it would be a feat. Little did it expect the two games to lure more than 1.5 million – each – within a year.

CrossFire's key to success came from its gameplay. Users joined online teams to complete goal-oriented tasks. Part of the fun comes from rising through the ranks from trainee to marshal. It also introduced different modes, including death-matches where teams compete to rack up the most kills, or elimination formats where players don't respawn.

Despite being free to play, people would pay to dress up characters or get better weapons. By 2015, it generated $6.8 billion in revenue, becoming one of the top-five highest-grossing games, along with Pac-Man and World of Warcraft. Even as of 2020, it was still the most played video game worldwide, with 6 million concurrent users, according to Smilegate.

The intellectual property rights for CrossFire have become so valuable that Fast & Furious producer Neal Moritz announced in 2015 he would produce a movie version of the game.

LEAGUE OF LEGENDS

A few years after Tencent's foray into hard-core gaming, it made its most important investment decision yet. In 2008, Tencent's gaze fell on a US gaming developer called Riot Games.

Riot was founded in 2006 by Brandon Breck and Marc Merill. The two college buddies studying business at the University of Southern California bonded over a shared love of video games and shared an apartment in West Hollywood after graduation. They pursued different careers – Beck as a management consultant at Bain & Co. and Merrill in business management.

Their fascination for games never abated. They would have back-to-back gaming nights in their apartment. They aspired to work on a title, but without any background in coding or design they struggled to find a viable path. The pair started by becoming advisers to the board of an online game company, helping the firm raise venture capital.

League of Legends emerged from a philosophy rather than a design document, Merrill told the *Washington Post* in an interview. They were inspired by a popular modification of Blizzard Entertainment's Warcraft III. The adjusted game featured two teams of heroes trying to take out the opposing side's base, but it was maintained by fans and lacked polish.

Modifications 'created this possibility space for people with relatively limited knowledge ... of various disciplines required to build a game,' Beck told NPR. People could 'make something powerful and special out of a lot of the existing assets of the existing game, and so there was a ton of experimentation and yeah there was magic in there.'

Breck and Merill saw an opportunity to take fan-tweaked games to the next level, creating a genre known as multiplayer online battle arena or MOBA.

In September 2006, the duo set up shop in an old converted machine shop under an Interstate 405 overpass in Santa Monica. Among the first interns they hired was a young student called Jeff Jew from the University of Southern California, their alma mater. They held a long chat about Warcraft and DOTA, and offered Jew an $11-an-hour internship.

Riot's office was 'super scrappy', Jew recalled years later, adding that the industrial, 1970s vibe in the building gave off an 'almost, like, wet feeling in the office.'[1] But the founders' philosophy was clear. They wanted 'gamers holding up their joysticks and keyboards, rioting for better games,' said Jew.

Riot started with a game called Onslaught, writing a big design document and focusing on creating characters. The goal was to create a demo in about four months for the 2007 Game Developers Conference in San Francisco. Given the lack of experience, the founders really underestimated the scope and complexity.

On top of that, Breck and Merrill hit a wall with potential publishers. The issue was rooted in a difference of opinion on the business model – they wanted to generate revenue from in-game sales instead of making a one-off profit from selling their product. That free-to-play model was something that Asian, and especially Chinese players, were familiar with. But in the US, such a notion was still in its infancy. North American publishers didn't want to take the risk.

The founders reasoned that a freemium model could lure more users and extend the life cycle of the game. Instead of having players make a one-off purchase, finish and discard the title, League of Legends would try to grow the game, expand its universe and extend its appeal via new characters, story lines and tasks.

The founders felt that in order for the game to succeed, it was vital that its publishing and development were completely aligned. That meant they had to go and raise money to fund their endeavour.

Fundraising was a daunting experience. They put together the essential fifteen-slide deck, with 200 extra pages as backup. Not giving up any chance to raise money, they reached out to everyone they knew.

At times, it was a thoroughly random experience. Through Merrill's girlfriend's dentist father, they met one of his patients, Greg Dollarhyde, the former CEO of Baja Fresh. They ran the pitch by Dollarhyde at his house in Malibu, who then turned to his seventeen-year-old son for

---

[1] Riot Games video

advice. After getting a nod, the businessman agreed to introduce more of his friends and network to the young founders for angel funding.

It was very much a 'living room roadshow', as the founders put it. In the end, they managed to raise seed money from about thirty individual investors, gathering about $1.5 million.

As they continued to travel around the country to talk to potential investors, they came across David Wallerstein at Tencent.

Wallerstein, the American who invested in Tencent on behalf of South Africa's Naspers, had relocated to San Francisco by now, helping the Chinese company scout for interesting startups. In 2007, he spotted Riot and kept in touch with the founders.

When Riot raised $7 million in 2008, Wallerstein talked to Tencent executives including Martin and Pony. He reasoned that while Riot had yet to produce a hit game, the company's MOBA concept seemed like great potential and the founders' philosophy of free-to-play games clicked with Tencent's.

Tencent by then had already embraced the philosophy of investing in a games studio, even though it was still largely a distributor for China. Its prime focus at the time was unearthing gems at good prices, licensing them for the local market to compete with other companies such as NetEase.

League of Legends was announced on 7 October 2008. Riot pulled out the stops, building out everything from maps, gameplay, characters, even virtual cosmetic items called skins that they hoped users would be willing to pay for.

Sales of skins and decorating avatars were concepts that had already taken off in Asia. It was the foundation of what is now known as micro-transactions in games, something that even Fortnite has adapted today. Tencent was a pioneer in such forms of payments. During the desktop era, it was largely enabled by its digital currency, Q coins. Users could buy top-up cards in convenience stores or transfer money to their accounts with Tencent via phone cards, and then spend the money in games. With the advent of online payments services like Alipay and WeChat Pay, small payments in-game have become even more convenient, especially for mobile titles.

But in the US back then, it was a nascent concept. Riot was also running behind schedule – a nerve-wracking experience, especially since it was racing against time to avoid running out of money – which prompted the company to hire more interns. Out of twelve designers, half were interns.

The competition was closing in, as others caught on to the idea of MOBA games. Members of the original DOTA were now working to create their own product, Heroes of Newerth. The teams raced to release their product within months of each other.

Riot came up with forty characters, and extra skins for half of them, right before the launch seven months later.

Despite Heroes of Newerth having better developers, art and more peak players upon initial release, League of Legends got one thing right: the business model.

League of Legends didn't charge to play the game, whereas Heroes of Newerth sold like a traditional title with a fixed price. After announcing its cost, Heroes lost half of its users, many switching to League of Legends instead. In hindsight, it was perhaps because many of the targeted players were in Asia and either couldn't afford the entry fee or were more accustomed to the freemium model.

In 2009, Riot earned an estimated $1.3 million in revenue. Income jumped to $85 million two years later. That prompted Tencent to take a serious look at acquiring a controlling stake in or taking over Riot.

Tencent executives – Pony, Martin and Mark – were all avid players of the title. Gaming was a way for them to do due diligence. 'We felt that you needed to look at the product. The product is a result of everything that goes in, and it tells you a lot about the team. The quality of the team, the culture of the team, the philosophy,' said Martin.

Riot's founders didn't just agree to the deal without conditions. They wanted to preserve the management structure, teams and culture of the company. Tencent's executives assured them Riot would be allowed to operate very much like an independent company.

That same year, Tencent took 94 per cent of Riot for $400 million. Tencent took two of the five board seats at Riot, leaving one seat vacant after the pair of founders occupied the other two. Riot continued to be responsible for its own profit and losses. Information between the companies would be shared sparingly, and Tencent employees didn't have access to the source code for League of Legends.

The structure of the acquisition – acquire but not incorporate Riot – would set an important precedent for other mergers and acquisitions. In fact, it pre-dated Facebook by a year in establishing the model, as the US social media giant also made the same arrangement when it bought Instagram. It offered smaller companies a way to keep growing, as opposed to competing in a battle to the death with the larger tech giants. Though that didn't mean rifts wouldn't emerge.

In 2011, Tencent introduced the franchise to China, where foreign gaming companies aren't allowed to operate without a local partner. Riot won more than 32 million monthly users in the ensuing year, with the majority of them in the Asian country.

Tencent's investment into Riot proved one of its best bets. It became a money-spinner. Riot opened offices in more than twenty countries, from London to Istanbul and South Korea. In 2015, it set up a new campus in West Los Angeles, signing a fifteen-year lease for about $16 million a year.

The campus offered Facebook-like perks with a free cafeteria, a movie theatre and meditation rooms. A falconer frequents the site to shoo off pigeons. The company filled a basketball court with real snow for a Christmas party, tech media outlet The Information reported, adding that Riot spent more than $5 million flying staff and family to the Dominican Republic or Seoul for annual gatherings.

League of Legends commands a following like no other, evident from the enthusiasm it generated during Tencent's Covid-era e-sports event in 2020. The game made $1.75 billion in revenue for the year, maintaining its status as one of the most lucrative titles globally. In 2021, *Arcane*, an anime series based on its fantasy universe, premiered on Netflix and Amazon's Twitch. It garnered 130 million views in China on Tencent's video site within a few hours, becoming the most-searched item on the platform.

While Tencent largely kept its promise to Riot over the years, the two had their challenges. Riot's fans – as the name suggests – have no qualms lambasting Tencent for being greedy and egging on players to pay more for in-game upgrades.

Money was also an issue. With Tencent as the distributor of the game in China, it kept at least 70 per cent of the revenue for the region. This became a source of frustration for Riot, as the country was its biggest market. Riot's staff felt they had given up too much.

Around 2014, Riot grew dissatisfied with the way Tencent was publishing League of Legends in China. The Los Angeles studio felt Tencent didn't spend enough time marketing the game, limiting its earnings potential. The US company quietly opened its own office in Hong Kong, exploring ways to exert greater control over the distribution of its most profitable franchise, The Information reported. It even hired a former Tencent executive who once oversaw a professional e-sports league to oversee publishing – then warned employees not to divulge the move to its biggest investor.

Unsurprisingly, Tencent wasn't amused when it found out. It refused to work with the new employee, who was supposed to act as a liaison. The Chinese giant would eventually end the minor revolt by buying the remaining 7 per cent stake in Riot in 2015. But its investee's stealth move opened fissures in the relationship that only widened when Tencent tried to make the transition to the mobile era.

FORTNITE BATTLE ROYALE

The investment in Riot paid off for Tencent not just financially. This desktop-era alliance with Riot opened up a plethora of opportunities in other companies. One of the largest and most important was with Epic Games Inc.

Tencent in 2013 forked out over $330 million for a 40 per cent stake in Epic, gaining access not just to one of the world's most popular and effective game-rendering engines, known as Unreal, but also to the mega blockbuster that would become known as Fortnite.

Epic's Fortnite Battle Royale has enthralled an estimated 350 million people around the world, gluing teenagers, rappers, professional athletes and middle-age accountants to their desktops, consoles and mobile devices. The company, which was valued at nearly $29 billion in April 2021, is the world's sixth most valuable startup, based on CB Insight's estimates at the time.

The premise of its flagship title is similar to that of Battle Royale's, a Japanese film about a group of students stranded on a remote island who're forced to battle to the death. In Fortnite, people either go solo or join a squad of players, with up to a hundred participating in every round. The game starts with people skydiving from floating buses and then scavenging for weapons and shields to defend themselves. As in the movie, the primary goal is to be the last player or team alive.

The game has become such a phenomenon that, in a 2019 investor letter, Netflix said: 'we compete with (and lose to) Fortnite more than HBO' for screen time.

Epic in many ways aligns with the vision that Tencent holds for its future. Its founder Tim Sweeney draws respect from the company's top echelons for his philosophy of combining gaming with entertainment (including movies), and for his musings on how virtual reality could hold the key to the future, in a sort of VR-based internet that's come to be known as the metaverse.

The low-key, normally media-shy Sweeney has lately made a name for himself battling big tech corporations, championing a crusade against Apple and Google over app store revenue splits. People who know him say he sees himself as fighting for the common man. While his worth stands at nearly $10 billion, he still wears t-shirts and cargo pants, drives a 2019 Corvette, and favours Burger King over fine dining.

Sweeney's journey to success began in his parents' garage. Born in 1970, he was raised in Potomac, Maryland. He had two older brothers. His father worked as a cartographer for the government, while his mother took care of them at home.

As a kid, Sweeney played Super Mario and visited his eldest brother in San Diego, California, at a startup he was working at. His brother taught him how to program on the company's IBM computer, introducing him to the world of coding. At eleven, Sweeney spent his time dissecting lawnmowers, radios and TVs. He immersed himself in the digital world via the Apple II Plus computer his brother gave him and used it to create video games. Sweeney later told the *Wall Street Journal* he spent more time 'programming than I think I was sleeping or in school or doing any other one thing in the world'.

To earn money, he offered to trim neighbours' lawns, undercutting professional landscapers when he was fifteen.

While studying mechanical engineering at the University of Maryland, he designed his first video game, ZZT, during his sophomore year. Because he had no idea how to program graphics, he used symbols and smiley faces as avatars to attack targets. The game would get harder as they upgraded in levels. He also founded his first company, Potomac Computer Systems – the entity that would later become Epic Games.

Sweeney dropped out of university, missing just one credit before graduation. He moved back in with his parents at the age of twenty. There he continued to work on his game, finally releasing it in 1991. He would find customers online and send them floppy disks of the game in return for cheques sent to his parents' house.

To beef up his credibility, he rebranded his company Epic Mega-Games, a name Sweeney said was 'kind of a scam to make it look like we were a big company'.[2] It worked. He sold several thousand copies of ZZT, which helped him move out of his parents' house.

Sweeney eventually dropped the Mega from Epic's name and moved his company to Cary, North Carolina, where it remains to this day. There, he helped program the company's inaugural first-person shooter game, Unreal, a PC-based online multiplayer shoot-'em-up. The 3D-graphics technology behind the game that helped render the images – adding

---

[2] The life and rise of Tim Sweeney, the billionaire CEO behind "Fortnite" who's now taking on Apple in a lawsuit that could have huge implications for the whole industry. (2020, August 19). Business Insider Nederland.

shading, colour and illumination to a two-dimensional image – was called the Unreal Engine.

The Unreal Engine became the foundation of Epic's success. It helped the company create its first breakout hit, Gears of War, in 2006. The *New York Times* described it at the time as one of the 'best-looking' games on the market. Gears of War flourished and became the face of the company, but it wouldn't be for another decade until the company would hit the big time again with Fortnite.

In the meantime, its rendering technology tided Sweeney over. Epic opened up the technology and offered it to other gaming companies for a fee, which became a main source of income.

'If we didn't have the engine, we would have died. We would have died three times,' Sweeney told gaming website Polygon, foreseeing how the Unreal Engine would be adopted in fields beyond gaming, including architecture, vehicle design and film creation. 'Ninety-five per cent of all these different applications' needs are the same thing,' he said. 'They all need photorealistic graphics, rendering, environments and then a great authoring tool and pipeline to link them all together.'

Even back then, Epic realised the potential that games could generate for cross-platform intellectual property, where games could be converted into anime, movies and vice versa. 'There's plenty of room for licenced Hollywood properties as long as the games are good and the properties would make an enjoyable video game experience,' Cliff Bleszinski, Epic Games' lead designer, told Bloomberg at the time.

The Unreal Engine earned a reputation for being state-of-the-art, catching Tencent's eye. Around 2011, Sweeney weighed selling some of Epic's shares as the company sought expansion and some early employees wanted to cash out. Morgan Stanley reached out to Tencent on behalf of the studio. Sensing an opportunity not to be missed, Martin flew out to Durham, North Carolina, to meet with Sweeney in person. He brought along a fellow Goldman Sachs alumnus he'd brought on board earlier that year, James Mitchell.

Sweeney greeted the pair at his home, impressing them with muscle cars he'd collected (Sweeney has since given up his expensive hobby). He then offered to show them Epic's office. Because the cars were all two-seaters, one of the two Tencent executives had to drive himself in a separate car. That task fell to James.

An Oxford University graduate with an air of patrician reserve, James was a former managing director at Goldman Sachs who led the bank's communication, media and entertainment research team in New York. The last time he drove a stick shift was when he took his driving test in Hong Kong. He had little idea how to drive a high-performance sports vehicle. So he chugged along at 30 miles an hour, stalling repeatedly. Luckily, it was a Sunday and there was not a lot of traffic.

The trip to the office was worth it though. It was like a scene out of *Iron Man*. Even then, Epic already sported several green-screen motion capture studios, cutting-edge at the time at a game developer. Over lunch at a steak and barbecue place, Sweeney impressed Martin and James with his knowledge of gaming, especially in consoles – something Tencent wasn't familiar with. The industry was then abuzz with chatter of the convergence of high-definition games with mobile platforms. But Sweeney said it would take years before mobile phone graphics-processing units could match that of PCs and Consoles in visual fidelity. He knew what couldn't be done, an important factor that differentiated him from other founders. Martin and James agreed it was clear Sweeney was in a league of his own, reminding the duo very much of Elon Musk.

The business was also attractive. On top of Gears of War, Tencent was impressed by Epic's Unreal Engine and saw its potential to underpin the development of the entire gaming industry. But mostly they wanted to invest because of Sweeney and the team, standing at about 450 people at the time. It was right around then that Tencent identified gaming as a pillar of its business. The company was looking to grow in not just content, but also acquire talent and their skill sets for its greater ecosystem. It foresaw how the gaming industry was shifting from box sales to the software-as-a-service model. Tencent wanted to become a global powerhouse in gaming, realising that if it failed to bring certain types of games to China, it would lose its edge in the domestic market as well. Morgan Stanley had told Tencent there was another bidder. Tencent initially thought Microsoft could be interested – in which case, they concluded that there'd be only a slim chance of them winning the bid. But James quickly found out through his own sources that it was Warner Brothers Entertainment.

Tencent couldn't compete with Warner Brothers when it came to helping Epic with distribution. In fact, it wasn't even able to immediately provide any help beyond money. But Martin and James managed to convince Sweeney by promising him freedom and independence. The talks with Epic coincided with a broader shift in Tencent's acquisition strategy. The company was much more open to buying minority stakes and granting its investees autonomy, instead of working on everything itself and competing with everyone. In fact, Tencent's executives – Pony, Martin and James – made it a point to retain founders after acquisition. It was not only a matter of building goodwill for future deals, it was also because they wanted talent that could create products the world had never seen before – on a serial basis.

Sweeney liked the idea of a high degree of flexibility. He also liked having a shareholder who wouldn't bully him into churning out new games for the Gears of War franchise. With that in mind, he agreed to sell Tencent 4 per cent of Epic for $300 million in 2013. Little did Tencent know, Sweeney would be putting their word to the test soon.

Six months after Tencent invested in Epic, Sweeney told Tencent that he wanted to sell Gears of War to Microsoft because he wanted to focus on another initiative for the future. As one executive put it, 'he said he wanted to burn the boat. And he did.' It helped that Tencent only held a minority stake in Epic, so it didn't have to incorporate the gaming company's financials into its balance sheet. But it was still nerve-wracking, as much of Tencent's financial modelling for the investment was based on revenue from Gears of War. Many other investors would have fought the idea of selling the franchise. But Tencent kept its word – and left Sweeney alone.

Two years following that investment, Sweeney made a decision that turned heads again. He offered Unreal Engine for free in 2015. The idea behind it was that Epic could enlist more, smaller developers to use its service. In return, it only asked for royalty fees on any game that made more than $3,000 per quarter.

Epic first revealed that it was working on Fortnite around 2011. The title in its early years bore little resemblance to the Battle Royale smash it would become – it would be another six years before Epic finally hit pay-dirt.

Forntite started off as a 4 vs. 4 game with Zombies, incorporating elements of world-building similar to Minecraft. But its original dark and lifelike concept met with a lukewarm response – there were too many similar concepts on the market. So Epic switched to a more cartoonish visual style, creating the pop-cultural hit familiar to fans around the world today. It improved its design in world-building, scavenging, terrain expansion and cooperation. Those went on to become signature ingredients of the title.

Its much-needed breakthrough came in 2017, and from South Korea. In March of that year, a game called PlayerUnknown's Battlegrounds (PUBG) went on sale and managed to sell 13 million copies within six months, shattering PC records and surpassing best-sellers like Grand Theft Auto V and Dota 2 in terms of number of users.

The creator, Brendan Greene, hailed from Ireland and had been living on welfare just three years earlier. He had a chaotic career, designing websites and shooting wedding photos in Brazil after a failed marriage. He also played computer games a lot. So much so that he became bored of his collection and started tinkering with existing code to create his own custom versions.

With basic programming skills, he modified military-shooter Arma 2 into a survival game also inspired by the Japanese film Battle Royale. The premise was similar to Hunger Games: a hundred players parachute onto an abandoned island, scavenge for crossbows, guns and frying pans, then proceed to try and to kill each other – until they are the last person standing.

Greene teamed up with a South Korean game developer called Bluehole, now renamed Krafton Inc., and the game became a breakout success on a scale that was almost unheard-of in the industry. Of course Tencent wouldn't miss out. It later invested in the company and helped Krafton pull off one of the country's largest initial public offerings.

PUBG's success fired up Epic, which then incorporated the Battle Royale concept in Fortnite and released a new version in September 2017. That was a turning moment.

Fortnite's availability on mobile made it accessible to people beyond the typical gamer community. Almost half of its players were women in 2018, according to research firm Apptopia Inc. For rival games, like Call of Duty and Grand Theft Auto, the figure stood at more like one in three, or fewer.

Taking a page from Tencent, Epic made the game free to play. Even so, the game still generated $5.1 billion of revenue in 2020, according to Sweeney. The key to its commercial success came from micro, virtual transactions. Instead of asking people to pay $40 up front, it enticed gamers to pay for digital items: motifs, costumes, dance moves, skins to dress up their avatars, or season passes through which players unlock rewards by levelling up. This is something people are surprisingly willing to do. These digital accessories went for anywhere from a few dollars to nearly $20 each.

While the virtual items gave players little competitive advantage, nearly 69 per cent of Fortnite players spent money on in-game purchases, and 37 per cent of those spenders said it was the first time they had ever bought virtual goods in a game, according to a survey by researcher LendEDU.

Fortnite is very much a pioneer of the season pass, a reward system that grants players virtual coins to unlock in-game skins and emojis. People can obtain it by playing the game, but also via direct purchases. Fortnite generates much more money from the season passes than other companies. Its peers including PUBG make 10 per cent to 20 per cent from this mechanism. For Fortnite, it's about half.

That business model helped resolve a long-standing issue for developers in the gaming industry. Studios have always found it a challenge to drive sustainable revenue, in a business littered with the corpses of one-hit wonders. The biggest game makers in the US – known as triple-A developers – mostly asked customers to pay upfront to download titles onto their desktops and consoles until 2018. Mobile games were the first to introduce Americans to in-game purchases, with FarmVille and Clash of Clans credited for pioneering the trend. Today, so-called in-app purchases are the closest the industry has come to a viable, recurring revenue model.

In fact, virtual items became such a goldmine, they would precipitate a global conflict, pitting Epic against the world's largest company, Apple.

UNDERDOG VS APPLE

Apple takes as much as 30 per cent of the revenue developers get from paid apps, in-app-purchases and subscriptions. That's no chump change: users of Apple devices spent $72 billion on the App Store in 2020, with almost $22 billion of that going to the iPhone maker, according to consultancy Sensor Tower.

'They don't allow competing stores – like, imagine a town that only allowed a Target and disallowed any other stores from building,' Sweeney told Bloomberg TV's Emily Chang. 'I mean, that's totally un-American and uncompetitive. But that's exactly what Apple does in an absolute sense.'

Sweeney argued that companies shouldn't be forced to use Apple's payment system and introduced its own. In response, both Apple and Google, which runs the larger Play Store, removed Fortnite from their respective app platforms – cutting access to Fortnite for more than a billion customers.

Epic responded by filing a suit. To generate public support, it also created an advertisement that directly parodied Apple's iconic '1984' advertisement, which symbolised the Mac creator's (initial) underdog status and potential for disruption.

In August 2021, Apple fired back by saying Sweeney was in fact trying to force a special 'side' deal that would fundamentally upend how the App Store works. 'Having decided that it would rather enjoy the benefits of the App Store without paying for them, Epic has breached its contracts with Apple, using its own customers and Apple's users as leverage,' Apple said in a court filing.

Apple argued that the App Store's success is related to its review process, and privacy and safety rules. The company's payments system ensures seamless experience and protects customers from fraud.

The battle came to a head in September 2021, with US District Judge Yvonne Gonzalez Rogers mostly siding with Apple – though she said the company should allow app developers to point users to outside payment systems. As of March 2022, Epic was still appealing the ruling with the support of 35 states, civil society groups and Microsoft Corp.

Apple plans to keep Fortnite off its App Store until appeals are exhausted in its legal fight. That could take five years, Sweeney said.

The Epic boss isn't done with uprooting the business model for mobile app platforms; he's fighting another war on the desktop game store as well. In a bid to challenge the world's largest desktop game distributor, Steam, operated by Valve, Sweeney opened the Epic Games Store in 2018. He charged game developers much lower fees than competitors. It had an immediate impact on the industry. Just before launch, Valve got wind of Epic's plan and offered to share more money with its biggest developers.

'The 70/30 per cent split was a breakthrough more than a decade ago with the advent of Steam, the Apple App Store, and Google Play. But today, digital software stores have grown into a $25-billion-plus business worldwide across all platforms, yet the economies of scale have not benefited developers,' Sweeney said in an interview in 2018 with Game Informer. 'In our analysis, stores are marking up their costs 300 per cent to 400 per cent. We simply aim to give developers a better deal.'

Sweeney estimated that if Epic received 12 per cent of revenue and gave the other 88 per cent to developers, the game distribution business would still be profitable. For games developed with the Unreal Engine, Sweeney waived their fees. While its selection is limited, it has a few plans in the works to encourage more people to download from its site.

One of them is relying on influencers. Originally a one-time campaign, Epic offered content creators like streamers on YouTube and Twitch a cut of the fee. The influencers can share referral links and get credit for sales on the Epic Games Store if they include their identity tags. Each title has a different revenue share set by the developer, but the minimum rate is 5 per cent, Epic said.

It successfully poached developer Ubisoft, convincing the company to offer The Division 2 in its store. Even though Steam had listed the game on its site earlier, Ubisoft didn't go with that company in the end. It scored another win with publisher Deep Silver, which released Metro Exodus exclusively through Epic's store in February 2019, despite previously being available to pre-order on Steam.

The exclusive deals drew criticism as well. Some users complained that Epic was forcing them to choose between the new alternative and their go-to site, which had a larger variety of items. User experience on the store was also less satisfactory compared with Steam, which offered more selection and a vibrant community that incorporated gamer feedback, reviews, forums, broadcasting and cloud saves. Sweeney defended his decision, arguing that developers should be free to choose which distribution platforms they work with.

'Love us or hate us, we are certainly fostering economic competition between stores, out of a firm belief that this will ultimately benefit all developers and gamers,' he said on Twitter.

The battle could last for years, and in the meantime Sweeney is pondering his other pet project – conceiving the metaverse and building a global entertainment empire. Sweeney took a $250-million investment from Sony in July 2020, lauding a 'convergence of gaming, film and music.' The company is said to be considering starting an entertainment division focused on scripted video programming, The Information reported, citing people familiar with the matter without naming them. The division could develop Fortnite into a movie, after poaching senior executives from LucasFilm.

On that, Sweeney's vision dovetails with Tencent's, which is now seeking a bigger overseas footprint, particularly as uncertainty engulfs its home market thanks to Beijing's persistent attempts to root out gaming addiction by curbing play time and censoring content.

Some argue that the metaverse – a concept encompassing virtual reality, augmented reality and social media – is the future of the internet and online entertainment. Internally, Tencent's executives think that it could be at least five years before virtual reality technology improves to the point that it goes mainstream – the makings of the metaverse, in other words.

For now, the company will continue to explore the same VR-based technology – say, allowing people to scan their own faces and drop them into games as avatars, or do the same with a building and plonk it into a virtual world. Epic's Unreal Engine could play a pivotal role in that. Used in the making of Disney's *The Mandalorian*, the sophisticated platform could well underpin a broader VR universe someday. All these business prospects mean Tencent and Epic are working together more closely. Sweeney speaks to Tencent's David Wallerstein on a regular basis and to James once every few months. Tencent discusses with Epic on how it can help with the backend-server setup and has provided advice on network architecture. While it's hard to get Fortnite into China right now – Tencent needs to wait for regulators to grant a commercial publishing licence and Epic announced in November 2021 that Fortnite would exit China after conducting beta-testing – most of Tencent's top mobile games, including first-person shooter titles, in the pipeline now use Unreal Engine.

Even as Tencent and investees like Riot and Epic look towards the next big disruptive technology, they have to wage a fierce battle to safeguard their respective slices of the mobile arena. Smartphones remain a powerful platform – and will for many years to come.

When it came to seizing the opportunities of the mobile gaming industry, Tencent was well prepared. It also managed to make the leap on its own, instead of relying on acquisitions and investments in other companies. In fact, you could even say it was ahead of its global peers. Gaming, in the mobile age, is after all what Pony regards as Tencent's bread and butter.

CHAPTER 10

ERA OF MOBILE GAMES

In 2012, Tencent's WeChat was on a roll. It had 200 million users and was venturing beyond instant messaging, linking content like news articles onto the platform. In a speech Pony gave about making WeChat an open platform that year, one thing stood out: 'When it comes to the mobile internet, the first thing that can generate profit on a large scale is probably mobile games.'

Coincidentally, just a year earlier, the US recognised video games as a form of art legally, ruling that video games were a creative, intellectual and emotional form of expression. That invigorated people in the industry, boosting their confidence.

Behind the scenes, Pony and his executives were paranoid that its gaming business could be disrupted during the transition to the mobile internet age. In op-eds that appeared in the state-owned *People's Daily*, he famously said: 'When giants fall, their corpses remain warm for a while', referring to fallen titans like Kodak, Blackberry and Nokia. 'These were all bloody examples happening right around us. When they were giants, we were still little brothers. We see how just because the giants failed to catch the trend in the slightest fashion, they fell.

'Even though Tencent's valuation is large right now, we still need to be vigilant.'

In response to Pony's call for transitioning to mobile, Mark Ren – who was head of gaming by then – reasoned it was time for Tencent to develop games on its own instead of purely licensing and distributing titles. This became one of the most important new initiatives within Tencent. The vision was to create five titles that could leverage the traffic of QQ and WeChat and catapult Tencent's gaming operations into the mobile era. Pony and Mark placed their faith in a young man called Colin Yao Xiaoguang.

Within the gaming world, Colin was already very well known. In fact, he was among three top product managers scheduled to present to Tencent staff during a key internal conference in 2012, along with WeChat's Allen Zhang.

GAMING FANATIC

Colin was best known internally for creating a hit racing game called GKart that attracted more than 2 million users online at the same time.

A broad-built young man who exuded quiet confidence, Colin's obsession with computers started young. In ninth grade in 1993, he pleaded with his father for a computer. The devices cost a fortune back then – the equivalent of months of his family's living expenses – which was why his dad was reluctant. That didn't curb his enthusiasm. He'd go to his father's office on weekends and spend whole days on the computer. That piqued his dad's curiosity, and he wanted to figure out whether his son was putting his time to good use.

One night, after Colin finished his homework, his father walked into his room with a book on the BASIC programming language and quizzed his son on coding fundamentals. He was astonished by Yao's knowledge at such a young age. He realised his son had talent.

So that summer, his dad took him on his bike in scorching weather and visited a dozen computer shops in the city to find the best bargain.

In the end, they settled on a personal computer powered by the Intel i386 processor. An antique now, it was gold for Yao. He was so excited he barely slept the first night. He also became the first person in his school to own a computer, marking a turning point in his life.

On his seventeenth birthday, Colin invited two friends over to play Blizzard's Diablo for the first time. He was fascinated by the design and game play. That planted a seed in his heart.

'I was deeply moved by the art design in games at seventeen. That's when I was determined to develop games for a living,' said Yao. 'Little did I know, China had no such industry at the time.'

Colin immersed himself in programming in university, learning everything from BASIC to C++. He got so good that teachers in his school asked him to help write programs for software used on campus. And after a restless summer, he delivered and earned 2,000 yuan in return, a handsome amount of money for an undergrad. He used it to buy an MX200 chip and a modem.

His fascination for Diablo never abated. In fact, he liked it so much, he tried to create a similar demo as a tribute to the game, just for fun, using Visual Basic and DirectX 7.0, a popular Microsoft tool for game development. With that demo, he landed his first job.

The person who hired him was Pony's competitor Chen Tianqiao, the founder of Shanda. Chen was just a fledgling entrepreneur back then. After winning his first pot of gold – 500,000 yuan via trading stocks – he set up Shanda. His vision back then was to create a cute virtual dog in hopes of generating a popular anime character and also game. But due to a lack of experience and funds, the project never went anywhere.

Colin had joined the IT department of a state-backed power plant in 2000 after graduation, thanks to his parents' connections. It was a comfortable job, but he wanted to do more with his life and subsequently quit, a move strongly opposed by his parents. They couldn't imagine Colin making a living from games. His mum agreed on the condition he find a job offering five times his salary of 1,000 yuan, or about $120, at the time. She thought it would be a good ploy to deter him.

To her surprise, he did. The founder of a small gaming company liked his demo and offered him that exact amount. The job also brought him to the capital of Beijing, which amassed the best talent and education resources.

Apart from rent and meals, Colin spent a fifth of his salary learning new skills at Peking University and audited classes for free to keep up with the latest technology. From there, he ventured into the world of online gaming and also helped create a few small hits, making enough money to support his dream of setting up his own gaming studio.

Inspired by Diablo's Chinese name, his first game's title meant Dark Online. Little did he know this would also become one of the darkest periods in his life. The startup drained his and his two co-founders' savings. They spent months cooped up in a small room writing code, barely talking to anyone. Despite all the hard work, the game only attracted a mere 4,000 people on the day of its debut and quickly got forgotten in a sea of other titles.

Worse, due to the long hours in front of a computer, Colin had to undergo surgery – his left leg had been in pain for two years. It took two surgeries and a replacement of his left hip to fix the problem. The agonising experience helped galvanise Colin, preparing him for the challenges to come.

After recovering, Colin decided it was time to find a job in 2006 and this time Tencent beckoned. He didn't disappoint, creating GKart within two years. So, naturally, he became a prime candidate to lead the company's transition to the mobile age.

The task that Colin took on was daunting. Essentially, what Pony and Mark was asking of him was not only to create a hit game but also to coordinate with multiple other teams within Tencent, a company notorious for competition between different units. To achieve the kind of results they wanted, Colin would have to work across Tencent's biggest business groups – encompassing fourteen different departments. Most of them had never collaborated before. Worse, his experience lay solely in desktop gaming development. Mobile games were a completely different beast.

SEVEN SINS

To start off, Colin took over a team focused on mobile gaming research, even though the staff had no experience in product development. He renamed the team 'Timi' – the anglicised form of its Chinese name 'Tian Mei', which reflected Colin's dream of bringing joy to people every day. It was allowed to operate as a standalone gaming studio, taking on more responsibilities for its own decisions but also for its profit and loss.

That turned out to be a great incentive. One time, Colin bumped into Tencent chief technology officer Tony Zhang, who asked him whether his team was working hard enough. Zhang half-jokingly pointed out that within the first eleven weeks of WeChat's inception, no one left the office before midnight. Colin smiled politely and said that no one on his team had left before midnight either – since day one.

It wasn't that they didn't want to, the workload just demanded it. Colin's team had five months to come up with its first batch of games, and it needed to coordinate with the powerful WeChat division, which often made last-minute requests and changes. WeChat was militant when it came to user protections. They were paranoid that users would be spammed by unnecessary ads and notifications, which could ruin the social media platform's reputation. That paranoia fostered uncertainty: often, the WeChat team would agree to grant access to certain software and then renege.

To their dismay, on the eve of the debut of their game, Colin's group discovered that WeChat's own internal team was racing against them to create so-called lite games. Allen Zhang's WeChat came up with a game that didn't require users to download and could be directly accessed within the messaging app's user interface. The idea was that of a casual shooter, where users manoeuvre an aircraft to shoot targets and dodge obstacles. It was a massive success and stole Timi's thunder.

Colin was gracious enough, at least on the surface. 'We need to thank Allen. Shooting planes helped nurture gamer habits of competing against each other, and it built a good foundation for the games Timi would create,' he told local reporters.

The competition didn't just come from colleagues. Tencent's habit of allowing different teams to compete internally meant multiple groups were racing towards the age of mobile gaming – regardless of top-down directives.

After Colin's studio acquired three smaller internal ones, they spent seven months creating a mobile game similar to League of Legends. The clock was ticking because they found out that a rival studio within Tencent, Lightspeed & Quantum, was working on an alike product. They began beta-testing their games almost at the same time.

The debut in the late summer of 2015 was sub-optimal. In fact, it was so bad Colin got snubbed by other departments when he asked for support. One key flaw was that the preliminary version seemed to have lost the feel of a multiplayer online battle-arena (MOBA) game – the essence of Riot's success with League of Legends.

At 9 a.m., Colin's team anxiously watched charts of user downloads. Instead of a nice uptick, what they saw was a lukewarm response and an influx of negative feedback.

Timi designed the game in a way that required users to hone their skills first to build up defence and attack power. That took away a lot of the excitement of the original, where people were accustomed to jumping straight into the action, hacking and slashing through an arena to tear down rival towers.

To shake things up, Colin took a deep dive into the issues within his team. He concluded that there were 'seven sins' plaguing the studio: cronyism – ten out of the eleven team leaders came from the same former company – and cliques; managers only delegating and not getting their hands dirty on actual work; too many executives with inflated titles; people getting promoted beyond their competency; staff unable to adapt to the Tencent culture and being inefficient; a lack of quality control, which became evident in everything from designs to operation; and finally, complacency.

VIRTUAL MERCENARIES

Colin ordered his team of 100 people to revamp their design. They made the important decision of doing away with the need to train skills and allowed players to get straight to battling. They changed the combat mode to five-on-five instead of three versus three, staying truthful to the MOBA game's original concept – even though that meant it required more character design, more computing power and more bandwidth – all while users demanded the same level of smoothness on a mobile app as on their desktops. They even changed the name to Honour of Kings in hopes of getting a new start.

The revamped title officially launched in November 2015. This time, it worked. For game makers, a few key metrics are make-or break benchmarks – user retention rate within a day, three days and seven days. If a title can maintain half of the users by the second day, 30 per cent by day three and 25 per cent within a week, then it is considered a success. Honour of Kings blew competitors out of the water. Even by day 30, it had a retention rate of 55.9 per cent. By August, it was projected to make $3 billion in its debut year and contribute more than half of Tencent's smartphone income that year.

It set a record in particular for one then-unusual metric: the highest proportion of female players for a hard-core or battle-arena title, genres dominated by males. For decades, game makers have struggled to rope in the long-neglected constituency and unlock their wallets. Females accounted for 54.1 per cent of users as of May 2017, according to internet consultancy Aurora Mobile. That outstripped the 35-per-cent-or-lower average for similar titles on computers or consoles.

In fact, Honour of Kings became so successful it spawned a black market for virtual mercenaries.

Because the company allowed users to link their ranking in the games with their WeChat accounts, that spurred something of a class system after the game went mainstream. With more than 200 million players hooked, mums, students and office workers started sussing out and sharing their colleagues' and friends' ratings. That summer, I found long-lost high-school classmates reaching out to me on WeChat and inviting me to gang up with them on missions. When I visited friends in Beijing, we'd pull out our phones at the end of dinner and battle away while waiting for our bills to arrive.

The game took around twenty minutes per session. Friends of five could form a group and hack and slash their way through a virtual arena to take down towers and the home base of opponents. It was tailor-made for smartphones, where people demanded fast-paced thrills and social media flair.

It became a race to impress – and many paid for status. I soon discovered that some of my friends had impressive rankings that clearly surpassed their fundamental skills. After prodding, they admitted they had resorted to paying professional grinders to help them climb both social and gaming ladders.

One such gun for hire was Huang Zhibin. The twenty-six-year-old helped almost two hundred people boost their rankings in less than a year. By day, he worked as a high-speed rail mechanic in the coastal province of Fujian. By night, he charged a one-time fee of about 2,000 yuan ($299) to help newbies gain a coveted 'Supreme King' label within a week.

Business was booming. On Monday night, he would be a thirty-year-old Shenzhen business owner slaying foes into the wee hours. Then he became a bloodthirsty university student from Shenzhen, and later a twenty-five-year-old office lady turned online warrior in Beijing. The side hustle helped Huang make about $1,500 a month – more than double his regular salary. 'These people have busy daytime jobs, and not all of them have the skills, but no one wants to be laughed at,' Huang said.

While some clients sought private tutoring from Huang, others wanted him to play on their behalf. The latter group willingly handed over their WeChat profiles so Huang could temporarily assume their identities. That's no casual sacrifice: WeChat is the first resort for millions of people when booking rides, ordering food or just looking up news, and comes with a digital wallet and entire contact list. But to many it was a small price to pay and Huang says their private information was of no value or interest to him. For those worried about money stored on WeChat, Huang advised them to set up payment passwords so he couldn't touch their funds.

One of his clients was Jewin Zhu, a thirty-six-year-old Chinese property developer based in Sydney who sought out Huang's services because he was unable to master the 'assassin' avatar he favoured. Yet Jewin felt he needed to. For him to maintain his social circle, the game served as an ice-breaker or bonding exercise for newly forged WeChat acquaintances. 'The main reason I play this game is purely for the purpose of social networking,' said Zhu, who owned at least six accounts so he had the appropriate ones to match his business contacts. 'When you meet someone, this is the easiest way to bridge a connection.'

Virtual mercenaries create a risk for Tencent. Honour of Kings itself is free to download and play. Tencent gets its revenue from users paying to upgrade their powers. The risk is that instead of buying a new amulet or avatar from Tencent to gain the upper hand in battles, that money would flow into the black market instead. It could also tamper with player experience.

Most game developers knew how much of a threat this posed. Activision Blizzard Inc., for example, bans trade in World of Warcraft accounts for fear of jeopardising not just the user experience but their own pricing power. Honour of Kings implemented similar policies, freezing or revoking peoples' accounts for violations.

But Tencent soon had bigger headaches than just cheating players.

RIOT RIFT

Honour of King's runaway success triggered an uproar at Riot's headquarters in Los Angeles.

The original creator of the MOBA games had missed the opportunity to jump on the smartphone bandwagon because the US company's leadership wasn't interested in making a mobile version at the time.

As hardcore gamers, Riot's team reasoned that the League of Legends experience couldn't be replicated on smartphones. They also didn't want to divert resources to mobile when the company was experiencing rapid growth on PCs, according to staff.

In 2017, I was told by staff that Timi Studios was preparing to roll out Honour of Kings to Western Europe and the US, attacking Riot directly on its home turf. It was a matter of great importance for Tencent. This was the first time the company was taking a creation of its own – one that had become the world's most profitable mobile game – beyond China. It was finally ready to establish its status on a global arena, as not just a distributor and passive stake investor, but also a bona fide content creator.

Tencent was ambitious, with plans to launch the overseas version of Honour of Kings, known as Arena of Valor, across the globe, including in Turkey, Thailand, France, Italy, Spain and Germany. To lure overseas players, it spared no expense, tying up with Warner Bros Interactive and DC Entertainment to allow players to assume the mantle of Van Helsing or Batman, for instance.

The relationship was further strained when Tencent used high-profile gamers of League of Legends to promote Arena of Valor, Reuters reported, citing people who declined to be named. To repair the damage, Tencent agreed to a two-month freeze in marketing for Arena of Valor after Riot complained to senior Tencent executives. Riot was later granted the right to veto Tencent's use of certain celebrity gamers and even marketing materials for the game, the people told Reuters.

Riot finally released its own mobile version of League of Legends in October 2021. But by then much of its thunder had been stolen, especially in China. Wild Rift only ranked fifth among the top MOBA mobile games in the third quarter of that year, according to mobile app researcher Sensor Tower.

Tencent and Riot's team, however, did manage to resolve their differences. Riot's executives told The Information that relationships with Tencent remained solid, while CEO Nicolo Laurent said that Tencent never formally asked Riot to make a mobile version of League of Legends. Perhaps both saw too much at stake not to repair rifts, especially since Tencent could help Riot break into the world's biggest gaming market, which also happened to have the toughest regulatory regime.

CLASH OF CLANS

Tencent also needed the partnership with Riot to work in order to keep other startups and companies open to taking its money. Those included the maker of the hit mobile game Clash of Clans. In 2016, Tencent led the $8.6-billion takeover of Supercell, getting its hands on one of the industry's most popular mobile titles through its largest-ever overseas deal. It led a consortium that bought 84 per cent of the Finnish company, valuing the gaming house at about $10.2 billion.

Chart-topper Clash of Clans was a strategy game tasking people to build their own village using resources they produce or gain from rewards, purchases of medals or even attacking other players. Contestants can band into clans, participate in clan wars and chat with each other.

It wasn't just the game that made Supercell a tempting buy. Founded by Mikko Kodisoja and Ilkka Paananen in 2010, the Finnish company was a collective of small teams, or 'cells'. The studio was built on the notion that a collection of these cells would produce a company with high efficiency and zero bureaucracy.

The company got its initial funding from the founders and borrowed from the Finnish government's technology funding arm, Tekes. Setting up its first office in Espoo, the second-largest city in Finland, the team of fifteen squeezed into one 30-square-metre room with six desks from a recycling centre and a single coffee-maker.

It was about a year into the operation that the company had to make its hardest decision. The team had started out to create games across platforms, including web games for Facebook, but by 2011 they sensed that mobile platforms like the iPad were the future. It took some serious internal debate and determination, but Supercell mustered the resolve to kill off all ongoing productions that weren't related to the iPad, and later mobile.

The change was especially hard to swallow for one team that had been going all-out to finish an-yet-untitled project, codenamed Magic. The five-member cell had worked day and night for nearly six months and was on the cusp of creating what they thought would be something amazing. Yet forcing the pivot towards mobile ended up being the right call: that particular squad or cell ultimately developed Clash of Clans, which, funnily enough, was codenamed Magic before its official launch.

In the process of producing mobile games, Supercell learned a lot from its failures. One of the most important lessons was that it's usually better to kill games earlier rather than later. The company introduced a key mechanism it dubbed 'company playable,' to ensure only the best games make it to market. Individual teams release their games to everyone in the company to test, with only the best passing this step.

'If it starts to feel like the game isn't going to work or isn't fun enough, it's usually a sign that you should have already killed it,' the Clash of Clans development team concluded.

Supercell wanted to make its marquee game accessible to the widest possible audience. They also prioritised making it as social as possible. When Clash of Clans launched globally in August 2012, they knew from the response that they had something special. It took three months for the title to become the top-grossing game in the US.

With success came capital. In 2013, SoftBank Corp., the telecom firm founded by billionaire Masayoshi Son, took control of Supercell in a deal that valued the Finnish studio with just 130 people at $3 billion. That meant each person created $23-million's worth of value on average.

Flush with cash, Supercell was inspired to pursue its long-gestating global ambitions, setting up offices in Japan and Korea. The team realised 'that for the first time in the history of gaming, it's now possible to create a truly global games company.'

But there was another tech giant watching the company's meteoric rise from afar: Tencent. Supercell's belief in a then-novel free-to-play business model also clicked with Tencent's. So when Martin Lau got wind that SoftBank was planning to offload its stake, he decided it was an opportunity he couldn't pass up.

Martin flew ten hours to Helsinki while battling a high fever, just so he could pitch to the company's senior executives in person. In preparation, he did what any top executive in his situation would: he played Clash Royale. A lot. So much, in fact, that he notched what at the time was the ninety-seventh highest score in the world. 'It's a way for me to do due diligence,' he says, with no discernible sarcasm. 'I still play it. But at that time, it was very intensive playing.'

When Ilkka Paananen, Supercell's CEO, heard about the score, he apparently expressed scepticism and challenged Martin to play against an accomplished employee. Martin won. Soon after, Tencent acquired a controlling interest in the Finnish studio.

That deal became one of Tencent's signature overseas acquisitions and it's paid off handsomely. As of May 2021, almost a decade after its launch, Clash of Clans is still growing strongly. Sales surged 79 per cent that month, while the international edition of the company's other title, Brawl Stars, jumped 64 per cent.

E-SPORTS CLUBS

With an army of titles under its belt, especially the in-house creation Honour of Kings, Tencent saw a golden opportunity to build an entertainment empire beyond gaming. The most obvious next step being e-sports, where players engage in competitive tournaments just like NBA players or Olympic athletes. With the success of Honour of Kings, a community of gamers mushroomed across China who streamed their sessions and strategy on social media platforms. Among them were many who aspired to become professional gamers who could compete in tournaments and even the Olympics one day.

Pony imparted the task of building a streaming and e-sports business to a different lieutenant this time: Edward Cheng Wu. A suave and intellectual-looking manager, Cheng honed his English during stints at Procter & Gamble Co. and Google.

When Cheng and I sat down, Tencent's e-sports event in Shanghai's futuristic Mercedes-Benz Arena was taking the nation by storm. More than eighteen thousand people packed into the UFO-shaped stadium just to catch a glimpse of their favourite players in live action. Like a modern-day gladiator mass spectacle, as many as 240 million people tuned in – more than double that of the Super Bowl – via their TVs, computers, cell phones and hundreds of game cafés across China.

On stage, the ten contestants were mostly in their late teens, with glasses. Their cherubic coach, no more than a few years older than them, fiercely jotted down messages on a board while blasting directives into his headphone, bearing a certain resemblance to Eric Cartman from South Park. Not far from the stage, parents of his teammates sat on nearby benches like chaperones at a school dance.

'E-sports is entering a golden age in China and globally,' Cheng told me. 'It's one of those few areas where China has a real chance of coming out on top to compete with developed countries.'

Those words became true in two years, when China overtook the US to become the biggest e-sports market in 2020, generating $21 billion in revenue. His ambitions, however, went far beyond professional gaming.

The core to the success of that strategy relied on an army of fledgling young gamers whose careers typically peak around their early twenties.

Tencent, with its dominant status in the gaming space, has often drawn criticism from parents and state media for inducing addiction. Stories about the abuse children suffer at treatment camps fuel even more of a public outcry and demand for curbs and accountability.

So I was intrigued to find out what separated a gaming addict and a top earner like twenty-three-year-old Zhang Yuchen. In 2018, Zhang was able to translate his obsession into a 10-million-yuan cheque when he switched clubs.

At typical gamer-retirement age, Zhang was still a top-flight player who led his country during the Asia Games in Jakarta in August that year. It's a responsibility writ large across his body language. He spoke into the floor beneath a mop of unkempt hair corkscrewed across his forehead – his red Adidas sneakers tapping against the carpet.

Zhang took a tough route towards gaming. Raised in Liaoning province – the pig-iron capital of China – he dropped out of school after tenth grade and became a roustabout with gigs including a realtor before turning his penchant for play into a profession. But whereas the games were light-hearted with flexible training when he started two years ago, teams today are typically on their headsets practising from noon until 5 the next morning.

'Maybe in the beginning we were only playing, but then we started to realise that in order to get results, you need to invest a lot and the training hours start getting longer,' said Zhang. 'It's counter-intuitive to the notion of playing games for relaxation or joy; professional gamers need to spend a huge amount of time thinking and studying.'

Much of what makes gaming beguiling and addictive is that the algorithms adjust based on who you are and what level you play, swaying users into a state of flow, where they fall under the spell of their own excellence – always just a grade below achieving perceived optimal performance – hence developing a constant yearning for improvement and reward, according to Nir Eyal, author of *Hooked: How to Build Habit-Forming Products*.

In tournaments, though, that illusion goes away. After Zhang was discovered by talent agents, his road to stardom began with lower-ranking matches.

Tencent designed e-sports with the goal of creating a professional network of clubs as serious as the NBA. Honour of Kings's intricate web of tournaments sustains a hierarchy of universities, cities and regions that, finally, rises to a national level. It spans the year with two finals for the spring and autumn season; and a championship, its equivalent of the FIFA World Cup.

PROFESSIONAL CLUBS

On-screen, Honour of Kings looks like a constant clash of light and movement. The game is extremely fast-paced, typically lasting only about twenty minutes a round. Players circle key choke points with millimetre precision, waiting for the perfect moment. Teams have to make a key strategic decision – to act or react – every three to five seconds.

At Zhang's level, gaming has become boiler-plate and institutionalised. He spends up to 70 per cent of his time watching and studying his competitors. To be a good captain, he needs to learn about the powers of all the characters, including tanks, warriors, assassins, mages and marksmen.

For every round that Zhang plays, there might be ten or more people supporting him – the clubs fly in coaches, data analysts, body trainers and psychologists to help maximise performance. Minor tweaks in a game can have massive effects on performance. Honour of Kings developers constantly adjust design and introduce variables to keep users engaged. That translates into alterations in the game's maps, ammunition and skill metrics for up to twenty characters at a time. Zhang needs to keep on top of all those changes. To him, the ability to feel the flow of the game and ride the wave at the right time – sensing when the game will spawn creeps and monsters that affect the buff (a gaming term for enhanced powers) for the entire team – is almost like poetry or art.

'Any small change has an effect,' said Zhang. 'Everyone is continuously studying; even very experienced people will face situations where they need to improve or suddenly have to start all over again.'

Investors saw the burgeoning opportunity. They poured money into at least ten thousand teams across the country in 2018, despite just twelve spots being available in Tencent's marquee King Pro League tournament.

The seriousness of the business – comparable to, say, the UEFA Euro Championship – meant investors in the top e-sports clubs were pouring in at least 20 million yuan of investment a year.

And with the International Olympic Committee considering incorporating e-sports in its ever-expanding roster, e-sports was garnering even more excitement.

Under the system, thousands of clubs vie for twelve spots at the top-tier King Pro League. If a team gets downgraded, then all that investment goes up in flames. The new system eliminates concerns for investors and club owners, said Allan Zhang Yijia, president of KPL and a general manager of mobile e-sports at Tencent.

To understand how serious the e-sports business in China can get, look no further than when the country won the League of Legends World Championship. A replay of the match was watched more than 15 million times within a week on the Tencent-backed video service Bilibili. Across the country, Chinese supporters erupted in cheers in college campuses, bars. They shouted words of joy from their building windows, waving flags, chanting and even streaking. The videos of the celebrations caused such a stir on social media that it drew ire from China's Communist Party Youth League, admonishing fans to react in a more rational way.

ENTERTAINMENT EMPIRE

Beyond e-sports, Cheng Wu had greater ambitions for Tencent's entertainment business. What he envisioned was a Marvel-like universe where Tencent could convert its intellectual property in anime and books into movies, games – and vice versa.

Cheng found his inspiration while recovering from back surgery – for spinal problems that stemmed from long hours in front of his desk and computer. One of the books he came across during his convalescence was called *The Tibet Code*, a 2008 adventure novel about an expert on the Tibetan mastiff, which later blossomed into a ten-book series that's been called an amalgam of *The Da Vinci Code* and 'Harry Potter'.

As of 2016, Tencent had already bought the rights to at least three hundred Japanese anime franchises in a push to become a worldwide multimedia brand like DC and Disney. Among the most famous was a blue-eyed ninja with spiky blond hair, Naruto. It also cultivated an audience of hundreds of millions of users who came to its platforms to read, watch and play games.

Cheng also went on the hunt for potential blockbuster targets in Hollywood and companies on both the creative and production side of movie-making. Some of Tencent's highest-profile film investments included *Warcraft, Men in Black: International, Venom, Kong: Skull Island* and *Top Gun: Maverick*.

'We will provide more of our input when we work with Hollywood partners,' Cheng said, adding that he sees franchises – in the science-fiction, comic-book and adventure genres – as a natural fit for Tencent Pictures and Chinese moviegoers.

MUSIC STREAMING

The larger entertainment strategy doesn't end with film.

The company has created music streaming services similar to those of Spotify, revolving around the division spun off and listed separately in Hong Kong known as Tencent Music Entertainment, which at its peak was worth $3 billion. The business, which many in China once dismissed as impossible to make money on, is profitable and generated north of a billion dollars of revenue in the three months through June 2021 alone.

Right before its much-anticipated listing, I sat down with Andy Ng, the company's vice president, to find out more about its plans to expand in a market known for copyright infringement and its strategy to fend off Spotify on its home turf – or even compete globally.

Amiable and avuncular, Andy is a Hong Kong native who worked for Nokia on music deals on the mainland before Tencent poached him in 2011. Back then, Tencent had yet to establish itself as a legitimate music distribution platform. In fact, the country was quite the backwater for professional content, rife with companies offering pirated music downloads.

A week into his job, he was called to an all-hands meeting. To his surprise, legal representatives joined the meeting. They handed him a list and his jaw dropped. There were at least twenty-four record labels suing

Tencent for infringement. 'My immediate thought was, did I make a mistake in coming here?' Andy said.

He soon showed that Tencent hired him for good reason. An old industry hand who had contacts at all the companies litigating, Andy called each of the labels, asking them to hold off on their lawsuits and give him two weeks to provide a solution. And within a fortnight, he did. He drafted up proposals on how to collaborate with music labels, the most important being to licence content from them legitimately.

The opportunity to sell his vision came not long into his job when he was asked to present big initiatives for the new year to Pony. He made one bold prediction: that the music distribution industry was on the cusp of a paradigm shift. Countries where piracy happened the most were also the first to leapfrog into subscription music services. Music piracy was rampant in Sweden and Korea, but they also were among the first countries to get users to pay. Andy thought the same could happen in China. As copyright lawsuits flourished, many companies would ultimately retreat from the business. It was going to cost them, but it was time for Tencent to go all-in.

'I said it's probably time for us to do something music-related at a time when everyone is giving up,' he recalled telling Tencent's leaders. But first, the company had to fix its mindset. 'If the piracy issue isn't fixed, then don't talk about a business model, don't talk about revenue.'

It was a big ask. Andy suggested that Pony should not even consider making money from music for at least four years. Pony didn't seem to mind, sending him on a contract-signing spree. Andy flew to Taiwan, South Korea and China trying to convince the big labels in those markets to work with Tencent. Within three months, he had successfully signed with at least eight, including HIM International Music, Jay Chou's JVR and Huayi Brothers. It wasn't easy.

'Like I said, Tencent, Alibaba, NetEase, Baidu, everyone had been infringing content for many, many years; you cannot all of a sudden say Tencent just washed its hands and said, "I want to become clean." The challenge that we were facing at the time from the music labels was, "Why should I trust you?"' said Andy.

But Tencent had a couple of tactical advantages. It was the first to reach out to the labels. And because of Andy's personal connections with executives, he garnered their trust more easily.

With the distribution rights, Tencent now had skin in the game. It had a reason to go after other local companies that were infringing on content it paid for. Andy quadrupled his team of lawyers to twenty and they sent out two thousand legal warnings alone in 2013.

In the following year, Tencent decided to make an example out of its fiercest competitor in the music business: NetEase. It launched a big lawsuit against the company, seeking compensation of 100,000 yuan per song. Tencent had exclusive music distribution rights with about twenty label companies by then. That represented an insane amount of money. NetEase was listed in the US and the news of them getting into such a brawl had the potential to tank its shares.

So NetEase's founder, William Ding, reached out to Pony with a peace offering. He proposed that NetEase make a token compensation to Tencent, then start sub-licensing the music rights from its rival, in effect establishing a model for the entire industry.

'My point of view was that if I could get one of the online portals to become legitimate, then the rest of the online portals would join the alliance,' said Andy.

His plan worked, and soon other players in the market including Alibaba reached out to sub-licence Tencent's content rights. Alibaba took down two million songs that infringed on copyrights within two years.

That also laid the foundation for Tencent's eventual dominance of the market.

With the issue of copyright out of the way, Tencent was finally ready to explore a viable business model. It looked at Spotify's, but concluded that in order to get Chinese customers to pay, it needed a more diverse range of content. That's when its signature app QQ Music offered subscription services that allowed users to download music to their devices, but also perks like priority access to concert tickets. In addition, QQ Music sold digital versions of single albums to lure more people into paying before gradually converting them to subscribers. Tencent's music business had 71 million paying users, nearly half of Spotify's, as of 2021.

Tencent Music's lordship over the Chinese market thwarted any ambitions Spotify might have had in the country. Also, the Swedish streaming giant simply lacked copyrights for local music, which accounted for 80 per cent of the local users' market preference. Yet Spotify was preparing for an initial public offering and investors were keen to know what its China strategy was. Tencent, on the other hand, had made few inroads into the global market, apart from limited success in Taiwan and Southeast Asia. That gave both companies a reason to work together.

With that in mind, talks ensued. Tencent's Martin spoke to Spotify founder Daniel Ek on the phone. James Mitchell, who by then had become Tencent's chief strategy officer, also conversed with Spotify's then-CFO Barry Mccarthy, who he knew from his past life as an analyst in the US for Goldman.

Barry suggested they have a meeting in the 'middle ground' of London in the late summer of 2017. It was an odd description as James had to take the longer journey and fly all the way from Hong Kong.

They convened at Goldman's headquarters, which turned out to be a short session. James got to London at 6 a.m. The talks started at 9 a.m. and were over by 9:07 a.m. Barry had experience with very tough negotiations from his time at Netflix when he dealt with studios. The Hollywood-style bargaining led Tencent to initially think that a deal couldn't happen, particularly after Spotify presented a take-it-or-leave-it offer.

James wasn't just ready to give up though. While travelling in Myanmar on a ship with no cell phone reception from Yangon to Mandalay, he had an epiphany on how to make it work. He came up with a proposal on a share-swap structure that could appease Spotify and make it work for both companies. Due for an investor meeting in New York the following week, he scheduled talks with Barry's deputy, Sheila Spence.

The two met three to four days in a row in New York to hammer out a deal, getting together either before 8 a.m. or at 6 p.m. to discuss and negotiate the right swap ratio. This time, it worked. In December that year, Tencent Music and Spotify announced they would invest in each other. Spotify would hold 8.9 per cent of the Chinese music streaming service, while Tencent and its music division would hold about 9 per cent collectively of the Swedish firm. That opened the gates to further collaboration, including sharing music content in respective markets. A year later, Tencent Music successfully listed in the US, raising about $1.1 billion in 2018.

But back then, Pony and his lieutenants had little time to celebrate their success. Tencent was about to run into its biggest challenge.

CHAPTER 11

STATE VS GOLIATH

Around April of 2018, Tencent staff got wind of a troubling policy change: that regulators were preparing to suspend approval of new gaming licences. It sent chills through them. It meant that the company wouldn't be able to generate money from its most popular games in the market, including the mobile version of PlayerUnknown's Battlegrounds (PUBG), which Tencent had only recently licenced to distribute in China.

Behind the scenes, regulators indeed froze game licences across the entire industry amid a government shake-up, following a reshuffling of power among different departments, sources told me at the time. The government was growing concerned about the prevalence of violence in games and how it was fostering addiction among the nation's youth. Officials even criticised Tencent for the growing incidence of myopia among children. It threw the world's biggest gaming market into disarray.

All games went through a two-step approval process in China back then. First, they needed to get registered with the Ministry of Culture and Tourism and checked for sensitive content. Then a second agency – which at the time was going through a restructuring and name change – decides whether to grant a licence and allow commercialisation. The agencies bar content they deem too violent, sexual or otherwise inappropriate. The US and Japanese governments, in contrast, let studios introduce titles on their own, while industry-backed trade groups bestow ratings based on language or violence.

Tencent got stuck with PUBG partway through this process. It had the first approval, so it let millions of customers download the mobile version of PUBG, but then it couldn't charge any money because it didn't have the second licence. There was a similar problem with Fortnite. It got approval to run trials with the desktop version, but it was also waiting for the right to monetise it.

While smaller competitors alerted investors in their earnings calls about the policy changes, global investors and funds with billions riding on Tencent's stock were late in cottoning on to the severity of the situation. Tencent's shares in New York tanked by as much as 10 per cent the day we reported the story. It also sent related Japanese gaming stocks like Nexon and Konami into a tailspin, as many companies counted on Tencent's distribution power to generate money for them from China.

The episode underscored the disconnect of the global investment community with on-the-ground developments with Chinese companies. That's despite Tencent having become one of the biggest holdings for global mutual funds, with numerous pension plans and billions of retirement dollars tied to its performance.

To many, Tencent was a proxy for China and the country's tech industry. In some ways, investors even betted on Tencent precisely on the faith that this company could navigate the maze of regulations in China. Up till the freeze on game licenses, it seemed like a sure bet, given the fundamentals: a billion users moving into the internet age, looking for online entertainment. What better place for your money, or so the argument went, than the company that commanded a monopoly on the country's gaming and social media scene.

Yet in the years I covered China tech, it always struck me how blind – or wilfully oblivious – analysts and investors were towards the policy risks lurking within the country's internet industry. Investors seldom engaged with policy-makers, more often analysing trends only at the sector and company level. Even after Tencent acknowledged that the government had frozen game approvals and that it was unable to generate revenue from PUBG, only one out of fifty-two analysts surveyed by Bloomberg imposed a negative outlook on the internet giant.

Congratulating management during earnings calls on their 'excellent results' and dismissing regulatory setbacks had become something of a norm. Folks in the financial community seemed unable to comprehend a day when the government would really bring the hammer down on its national champion, constricting a spigot creating companies that, for the first time, could rival those of Silicon Valley.

Little did they know how things would change.

That June quarter in 2018, Tencent's profit fell for the first time in at least a decade, due in part to its inability to profit from its most popular games, including PUBG. To appease Beijing, Tencent would have to go through drastic changes.

It started by limiting the time children could spend on its games. In September, Tencent began using the police database to verify the ages and identities of players on its biggest mobile title, Honour of Kings. Even for a nation whose internet users were growing ever more accustomed to surveillance, the move was still a first of its kind, a step up in policing user activity and revealing their identities.

Tencent flipped the concept of PUBG on its head. Instead of building a game on the premise of 100 people killing each other till there was one person standing, it created a copycat version of the game with a storyline about patriotic warriors in training drills, paying homage to the Chinese air force. The title, called Game for Peace (or Peacekeeper Elite, as some dubbed it), mimicked PUBG's gameplay, down to a very similar interface and functions to help users migrate their in-game profiles.

With typical Chinese ingenuity, Tencent scrubbed the game of violence, swapping green liquid for blood, inserting banners that proclaimed nationalistic slogans, and portraying game victims blithely waving upon their departure rather than expiring in sensationalistic fashion. In fact, the company asked the military's recruitment arm for advice during development. Five winners share the final glory, instead of one lone survivor.

That prompted an outcry from the gaming community. More than 80 per cent of reviews for the game on Apple's Chinese app store were negative during its first week, according to Sensor Tower. Gamers lashed out against awkward sound-effects, fawning Chinese patriotism and the whitewashing of violence. 'Even though it's playable, it feels so weird for us veteran gamers,' one online commentator wrote. 'Give me back the excitement,' another said. 'Trash. Delete,' wrote a third.

Most of all, they missed PUBG's tongue-in-cheek chicken dinner – awarded to the last person standing. That became frosted cake in Tencent's world.

But it worked. The government granted the company a licence to make money off the game, which was what Tencent wanted all along. Without a better substitute, Game of Peace earned Tencent $20 million in the first five days upon launching in May 2019. It finally helped catapult the company into the thick of a *Hunger Games*-style genre that took the industry by storm, jumpstarting the Chinese company's waning growth.

Then it hit another snag.

US INSPECTION

It took two decades and Trump winning an election, but the US finally started to catch on to the growing influence of China's tech industry. In June 2019, US Senator Marco Rubio directed his firepower on the billion-dollar money trail that links American funds with China's internet.

Specifically, he questioned MSCI Inc., asking the global index provider why it had added hundreds of Chinese stocks to its benchmark emerging markets index the previous year, spurring the outflow of billions of passive investment dollars towards Chinese companies. That's because the MSCI emerging markets index is tracked by more than $1.8 trillion of collective capital.

Rubio warned that the US needed to become much more careful about providing money to a country increasingly contending with

Washington for geopolitical leadership. 'We can no longer allow China's authoritarian government to reap the rewards of American and international capital markets,' he wrote.

'Firms like MSCI have an obligation to make sure investors know whether their investment dollars are unwittingly aiding Chinese state-owned and state-directed companies linked to China's efforts to steal American innovation, undermine fair competition, increase threats to US national security and economic security, and support China's systemic and egregious human rights abuses,' he said. MSCI pushed back against the criticism.

'Currently there is no US law or regulation that prohibits an index company from creating an index containing China's A [shares] or US investors from trading in the China [A shares] market,' MSCI's Chief Executive Officer Henry Fernandez said in a letter seen by Bloomberg.

In tandem with Rubio's comments, Donald Trump and his administration also upped scrutiny of ties between the US and Chinese technology companies such as Huawei Technologies Co. The US blacklisted the telecom equipment giant as a threat to national security, cutting it off from American suppliers. Huawei challenged the blacklisting, also asking a US court to overturn a Federal Communications Commission ruling that the Chinese telecoms equipment maker poses a security threat because of alleged ties to the Chinese military. It argued in a filing that the FCC's ruling was 'arbitrary, capricious' and not supported by substantial evidence.

It was the growing tension between two superpowers – the last ones left on the planet – that finally shone a spotlight on that money trail. The realisation, all of a sudden, dawned on Washington's power-brokers that China had become a force in 5G networking, fintech and artificial intelligence to rival the US.

In some ways, China had the US to thank for its unprecedented digital evolution. A large chunk of the capital behind its tech giants' success can in fact be traced back to funds that manage money for Texas teachers, San Francisco firefighters, Minnesota policemen and Louisiana judges.

The model has persisted for more than two decades. It works like this: pension funds from California to New Jersey and college endowments pour money into venture capital and private equity firms, which in turn scour the globe for the best investment opportunities. In China, they struck gold.

It surprises many that more than 90 per cent of US foundations and university endowments have some exposure to the Asian nation. That makes it harder to separate out China investments made through vehicles, such as index-linked funds that allocate part of the money to the country.

Further complicating the situation, US investors had staked a lot on Chinese startups. Tiger Global, one of the world's biggest investors in billion-dollar startups, has 10 per cent of its unicorns in China. Coatue Management LLC, also New York-based, has 22 per cent of its unicorn portfolio in the country, data provided by CB Insights in 2021 show. Goldman Sachs had 18 per cent.

American investors have in fact been one of the biggest beneficiaries of China's initial public offering boom. If a US investor subscribed to $10,000 of Baidu shares during its August 2005 IPO, that investment would have grown to more than $422,000 by June 2020: a forty-two-times return. A buy-in to Trip.com's 2003 IPO of $10,000 would have netted $1.2 million.

All told, the S&P/BNY Mellon China ARD Index rose 351-fold between 2004 and 2020, compared with a 195-fold rise for the S&P 500 index.

The model of raising American money to fund Chinese startups in the world's biggest internet market has worked successfully for two decades. Private equity and venture capital firms wanted this model to persist.

Amid all the tension, Tencent tried to stay low key; Pony Ma all but vanished from public events. Yet its sprawling business meant it was an easy target. Avoiding getting caught in the firing line between China and the US was almost impossible for a company of its size. All it took was a thorny issue, like Hong Kong.

In October 2019, a Twitter post by an NBA executive expressed support for Hong Kong pro-democracy protesters. That incensed Beijing and mobilised an army of online patriots, inciting a backlash that was enough to prompt China's largest broadcaster to suspend game telecasts. That left Tencent in jeopardy.

Tencent had just inked a $1.5-billion, five-year deal to stream NBA games online in China. Almost half a billion basketball aficionados tuned in during the previous season. Now it was forced to suspend broadcasts, following a similar move by the state broadcaster.

Another firestorm erupted after Blizzard, the World of Warcraft studio partly owned by Tencent, banned a gamer for endorsing the same pro-democracy movement in Hong Kong, triggering calls for a boycott to punish its apparent kowtowing to China.

Epic's founder, Tim Sweeney, tweeted his disapproval of Blizzard's action, again inspiring a wave of Chinese nationalists to shun its Fortnite blockbuster. The questions from China's angry netizens: 'Tencent, why aren't you holding your dogs on a lead. They are biting you in your face.' (As mentioned before, Tencent owns 40 per cent of Epic.)

Americans were equally fired up. From fourteen-year-old gamers to NBA Hall-of-Famer Shaquille O'Neal, people started asking how one Chinese company – Tencent – got to dictate what people in the US can and cannot say?

The pressure only increased. In May 2020, the US Senate over-whelmingly approved legislation that could lead to Chinese companies getting barred from listing on US stock exchanges. The bill, introduced by Senator John Kennedy, a Republican from Louisiana, and Chris Van Hollen, a Democrat from Maryland, passed with unanimous consent. It required companies to certify that they are not under the control of a foreign government.

At the heart of the issue was a problem that vexed US regulators for more than a decade: China's refusal to let inspectors from the Public Company Accounting Oversight Board review audits of mainland companies like Alibaba and Baidu that trade on American markets. The effort gained strength after an accounting scandal at high-profile US listee Luckin Coffee highlighted for regulators the risks to American investors inherent in spotty disclosures.

The bill dictated that if a company couldn't show it wasn't under foreign government control, or if the Board wasn't able to audit the company for three consecutive years to determine for itself that it wasn't, then the company's securities would be banned from American exchanges. While not technically part of government, the Board is overseen by the Securities and Exchange Commission.

It sent ripples through Chinese markets, hammering tech company shares in particular. The combined market cap for companies facing delisting threats amounted to more than $2 trillion at the time.

The big question: if thousands of startups in the country were unable to tap one of their most important capital-raising avenues, what was the alternative?

The most obvious answer was Hong Kong, an Asian financial centre regarded as one of the most important reasons behind China's economic miracle. Hong Kong has been the largest single source of capital bankrolling its rise, according to a study by the Hinrich Foundation. From 1985 to 2014, Hong Kong was the source of 47 per cent of the country's foreign direct investment. Scholars have argued that one important reason Russia stalled economically is that it lacked its own financial centre to link up with global capital.

Yet Hong Kong at the time was still in the throes of anti-government protests. A 2019 proposal to amend the city's extradition laws triggered years of social unrest that frequently brought the city to a standstill. It alarmed Beijing, prompting crackdowns that further led investors to question Hong Kong's long-term viability as an international financial centre.

The trouble didn't end there. Trump had his eyes squarely on the biggest players as well. Particularly Tencent's WeChat. In September 2020, Trump announced that WeChat would essentially be shut down in America because it was deemed a national security threat. The administration said the Chinese Communist Party has demonstrated the means and motive to use such apps to threaten security, foreign policy and the economy of the US. Tencent said at the time it was reviewing the commerce department statement and working with the government to come up with a long-term solution, declining to comment further.

Mark Zuckerberg was quick to jump on the bandwagon, after failing to woo the Chinese government and get Facebook into the country. In testimony to Congress, Republican representative Patrick McHenry asked the billionaire Facebook founder if Tencent's WeChat Pay is a competitor to Facebook, and why the company hasn't built its own version.

Zuckerberg answers: 'I think you're right that they're certainly competing, not just with us, but all of the American companies on this.'

The Chinese community – many of whom were WeChat devotees – in the US panicked. Many rushed to secure their app before the ban took effect. Downloads surged, making WeChat the 100th most-downloaded app in the US on 18 September, according to mobile analytics firm SensorTower. Before that, it typically ranked between 1,000th and 1,500th on the iPhone store.

Tencent faced the risk of further disruption in the US. Trump banned American companies from transacting with WeChat, meaning people still using the app could experience slowdowns and glitches because the third-party services that helped keep WeChat running couldn't operate as smoothly as before.

In an ironic twist, Trump's plan was vetoed by a magistrate judge in San Francisco. That upended his effort to halt the use of WeChat, when the judge issued an injunction at the request of a group of American WeChat users. They argued that the President's prohibitions would violate the free-speech rights of millions of Chinese-speaking Americans who rely on the app.

WeChat 'serves as a virtual public square for the Chinese-speaking and Chinese-American community in the United States and is (as a practical matter) their only means of communication,' the judge wrote in the ruling. The court found the government had provided insufficient evidence of a security threat.

The drastic turn of events surprised many Chinese people, who marvelled at how a magistrate could veto the head of state. 'If this were to happen in China, it would basically be the equivalent of a county official telling Xi Jinping to get lost. Can you even imagine something like this happening in China,' a Chinese manager working for an influential US fund told me. 'For the Chinese, this was a great demonstration of the balance of power in America, and shows that the US is a system that has strong self-correction mechanisms.'

It seemed like a failed, last-ditch effort on Trump's part to seal his legacy. He was due to hand over power in a few months, after losing the election.

The two moves by the US – threatening to toss Chinese companies off American bourses and supporting the Hong Kong pro-democracy movement – struck many in Beijing as sinister. If the two superpowers were playing 3D chess, it promised to derail a thriving tech sector that had created hundreds of thousands of jobs and billions in wealth over an unprecedented decade of expansion.

The Trump crackdowns prompted the Chinese government to step up, retorting that the US action against WeChat was 'economic bullying'. It gave some the impression that the government would finally ease its own campaign against Chinese internet companies at home, and once again embrace the free-wheeling industry.

And for a few months, it seemed that way.

CRUSHING FINTECH

When I talked with executives at China's largest internet companies in early 2020, some said – half-jokingly – that Trump had saved them from Beijing's iron fist. Just a year ago, right around the time of the campaign against addiction and Tencent's gaming business, internet companies were starting to sense a significant increase in scrutiny over their own operations. Now they thought they could finally catch a break back home.

With the US attacking Chinese companies and Hong Kong's status as an international financial centre in jeopardy, Alibaba founder Jack Ma saw a perfect opportunity to demonstrate his value to the country.

Starting around mid-2020, chatter began circulating that Jack was planning to list his crown jewel Ant Group, the company that owns Alipay. The fintech giant's float was expected to be the largest in the world. It took months of chasing, parsing inaccurate rumours floating around the financial community, but by July we had broken news that Ant was planning to list simultaneously in Hong Kong and China's new tech board in Shanghai.

It seemed like a brilliant move to curry favour from Beijing. On the one hand, Ant's IPO would bolster Hong Kong's status as a financial centre. On the other, it would be the first big significant listing on the so-called STAR market, a fledgling board personally championed by Xi Jinping.

Indeed, Ant's founders were immensely proud of their achievements. A senior executive I spoke to hailed how it marked 'a watershed moment in flow of global capital.' It was a far cry from the days when Alibaba's founders had to tour seven destinations – including Kansas City – to attract investors to their own IPO.

In Ant's case, the entire roadshow took only a matter of days, mostly via Zoom. More than 80 per cent of the money it raised came in from parts of the world, including China, the Middle East, Europe and Southeast Asia, while US investors only stumped up little more than 10 per cent.

The IPO was so sought-after that global investors beseeched Ant execs for allotments, trying to show they were valuable as long-term investors. In Hong Kong, nearly a fifth of the population was estimated to have signed up for Ant shares.

It was a crowning moment for Jack, who was on the brink of pulling off the unprecedented $35 billion IPO. It would have conferred upon the company he founded two decades ago a valuation of more than $300 billion, swelling Jack's personal fortune beyond $61 billion and cementing his position as the nation's richest man.

In fact, the triumph marked such a feat for the tech mogul that he proudly declared in October 2020 during the now-infamous Bund Summit in Shanghai that this was a 'most critical' time in the development of finance. In his moment of glory, the billionaire roasted anachronistic government regulation, warning that it would smother innovation in China.

It was a classic performance by the famously outspoken executive, known for his confident swagger and soaring rhetoric. But this time, he stepped over the line. What ensued can only be described as a coordinated attack on Jack's financial juggernaut.

First came the suspension of the world's largest IPO. On Monday morning, days before the listing, Jack was summoned to a conference room at the China Securities Regulatory Commission in Beijing.

When mid-level bureaucrats finally turned up, they skipped over pleasantries and delivered an ominous message: Ant's days of relaxed government oversight and minimal capital requirements were over. The meeting ended without a discussion of Ant's IPO, but it was a sign that things might not go as planned.

Jack's meeting triggered a behind-the-scenes scramble by Ant and its bankers for more clarity from Chinese regulators. While securities officials signalled at the time they weren't aware of any changes to the IPO plans, the regulator later published a cryptic social media post describing a 'supervisory interview' with Ma, setting tongues wagging from Hong Kong to New York.

By Tuesday afternoon, the mood had worsened as whispers of a delay began circulating in Shanghai. And at around 8 p.m., the city's stock exchange called Ant to say the IPO would be suspended.

When the official statement landed less than an hour later, it cited a 'significant change' in the regulatory environment but offered few additional details on why authorities would scupper the listing two days before shares were expected to start trading.

At a hastily arranged meeting between Ant's bankers and the CSRC later that evening, officials pointed to the company's need for more capital and new licences to comply with a spate of regulations for financial conglomerates that had begun taking effect at the start of November. There was no discussion of how quickly the IPO could be restarted.

In retrospect, the swiftness with which Beijing moved to pull the plug on Ant was a harbinger of a much broader – and perhaps more dangerous – shift taking place in the highest regulatory echelons.

Only a week later, the antitrust authority issued twenty-two pages of proposed anti-monopoly rules, which many read as a veiled warning to Jack and fellow entrepreneurs to tone down the swagger. As the attacks mounted, China's Politburo, the top decision-making body of the Communist Party, in December emphasised the need for stronger antitrust oversight and to prevent the 'disorderly expansion of capital' – a signal that private enterprises should expect stricter controls.

Investors watched in awe as China torpedoed the ambitions of its national champion. The bigger question was whether the crackdown on Ant was an isolated incident, or whether it signalled more trouble to come for the wider industry. By the end of 2020, many were still blithely betting it was the former.

SILENCING BILLIONAIRES

Trouble in the US was still brewing. In November 2020, the US again threatened to delist Chinese companies. The US Securities and Exchange Commission said it was weighing plans to kick Chinese companies off US stock exchanges by the end of the year, if they failed to comply with US auditing rules.

'All of a sudden, it felt like China's internet companies were pariahs,' one internet founder told me. 'It's like we're getting attacked both from the front and back, not getting a moment's reprieve.'

Over the following months, more details emerged as to why China decided to crush its highest-profile IPO. People started to realise that the derailment of Ant was only the beginning and it marked a watershed moment: China was preparing to draw the curtains on the golden age of tech in the world's largest internet market.

For years, the fintech industry presented regulators with significant challenges because of its huge client base and growing role in China's money flows and financial plumbing. Ant was the biggest player, hence it was first to suffer the blow. But in ensuing months, Tencent and a slew of smaller competitors would be reined in and ordered to overhaul their business.

China's new generation of fintech companies made financial services more convenient and accessible to hundreds of millions of users. But they were also destabilising the country's financial infrastructure and chipping away at the profits of state-owned banks – all vital for the Communist Party's power base.

In tandem with the crackdowns, China's other regulatory departments – including its anti-monopoly watchdog and cyber administration – went into overdrive.

New rules to curb monopolistic practices across the entire internet landscape were drafted and finalised in February 2021 over a span of three months. Regulators levied a $2.8 billion fine against Alibaba in April and accused the company of monopolistic conduct, ordering numerous 'rectifications' to how it does business.

The government also went after the online grocery units of companies for predatory pricing. Among them was the country's largest on-demand delivery platform, Meituan, and its second largest e-commerce company by market capitalisation, Pinduoduo.

The sweeping overhaul sent chills through China's richest men, especially those who obtained their fortunes during the roaring internet age. One by one, the country's tech billionaires began fading into the background, stepping down from chief executive or chairman posts. Many began donating to charity in a big way. Meituan's Wang Xing donated around $2.7 billion of stock to a personal charity promoting scientific research and education. Pinduoduo's Colin Huang gave $1.85 billion to an educational fund. He Xiangjian of the Midea home appliances empire forked over $975 million, and Hui Ka Yan, founder of the now fallen real-estate empire Evergrande, $370 million, to help with poverty alleviation and medical care.

Amid the crackdowns, interesting things were happening on the capital markets front. Financial markets were soaring thanks to post-Covid stimulus plans, hitting record highs and fuelling an IPO boom. Chinese startups, sensing that this might be their last window of opportunity to go public before markets turned south, went into overdrive, particularly in the US where they were still commanding higher valuations in general than in Hong Kong, and faced less stringent rules.

UNPRECEDENTED PUNISHMENT

Among them was China's largest ride-hailing company, Didi. Backed by Tencent, Cheng Wei's company opted to list in the US after witnessing how a smaller competitor ran into trouble going public in Hong Kong. For both companies, a significant proportion of their business operated in a legal grey zone. China placed strict restrictions on drivers, car models and licences, demanding that drivers on Didi's platform have local identity registration. But in order to meet the demands of customers, it was simply not realistic to only employ drivers with valid local IDs, especially in places like Beijing and Shanghai, where driving for Didi was sometimes deemed beneath local residents. The Hong Kong stock exchange typically asks companies not fully compliant with the law to obtain some form of official blessing from key regulators – an almost impossible ask.

In the US, the Securities & Exchange Commission doesn't require companies to be fully compliant with local laws in order to list. They only need to disclose the risks. That prompted Didi to bypass Hong Kong and seek a listing in New York instead, people familiar with the matter told me at the time.

Didi's US IPO looked at first like a great success, raising $4.4 billion in June 2021, the largest offering in history from a Chinese company after Alibaba's. It turned co-founder Cheng into a billionaire and rewarded long-time backers SoftBank Group Corp., Tiger Global Management and of course Tencent.

Then their worst nightmares came to pass. Chinese regulators pounced two days after Didi's debut, banning its service from the country's app stores. Turns out, Didi's journey to go public was far more complicated than it let on. The following details emerged from talks with people involved in the process, who spoke on background.

Didi began discussing IPO plans with its bankers at Goldman Sachs, Morgan Stanley and JPMorgan late in 2020. The company weighed up whether to list in Hong Kong or the US. At one point, a small number of its shares traded at a valuation of about $100 billion on secondary markets. That set a high bar for the company.

By March, they homed in on the US because the rules were more amenable and the company expected a better valuation from investors familiar with its American counterpart, Uber. That was after the Hong Kong exchange, as anticipated, questioned Didi's compliance with Chinese regulations. Many of its drivers lacked a local household registration, or hukou, for the cities where they lived, part of municipal requirements for providing on-demand ride-hailing services.

In the months leading up to the listing, China's regulators had largely supported the idea of an IPO for Didi. But others had expressed concerns about its data security practices since at least April. In one example of concern, Didi had publicly disclosed statistics on taxi trips taken by government officials – a powerful but alarming display of its data capabilities.

Regulators privately urged Didi to ensure the security of its data before proceeding with the IPO, or shift the location to Hong Kong or mainland China, where disclosure risks would be lower. Officials didn't explicitly forbid the company from going public in the US, but they felt certain Didi understood their instructions, said those au fait with the situation.

One person involved in the meetings asked why Didi didn't act on suggestions from regulators, and uttered a proverb to the effect that you can't wake a person pretending to sleep.

Whatever the case, Didi and its bankers raced ahead – and things got messy. As the company prepared to make its first public filing with the Securities & Exchange Commission in the US, its own bankers weren't sure when the documents would land. The filing ultimately hit about 3:45 a.m. China time on 11 June.

During that period, Didi's government relations team handled discussions with the cybersecurity watchdog, and management relayed the content of those talks to bankers. Didi knew the agency had concerns about its data practices, but executives didn't think officials there had vetoed the IPO, according to people involved in the process.

Didi faced the choice of erring on the side of caution or going ahead with an offering that would fill the company's coffers and enrich executives. On 28 June, Didi's management gave the green light.

The company told its bankers it was allowed to go public provided it keep a very low profile, one person familiar told bloomberg, adding that the bookrunners were told by the company that there would be no press release to announce the IPO. Didi didn't even publicise to its own employees its impending New York listing – a landmark for the still-young company – until the last minute. Near midnight on 30 June, the company posted an announcement on an internal forum, another person said.

On Thursday, 1 July, Didi's shares surged about 16 per cent, a sure sign of robust investor demand. By Friday, Didi's management began to relax and celebrate. It was time to pop the champagne.

Then came the bombshell. That evening, after 7 p.m. China time, the cybersecurity watchdog posted a notice on its website: the agency is launching an investigation into Didi to safeguard national security and protect the public interest. Didi didn't respond to requests for comments regarding the IPO process and investigations at the time.

The fallout was extraordinary. Didi's share price tanked, shedding nearly half its value in a month. In December, just five months after going public, Didi announced it would begin preparations to withdraw from US stock exchanges, an unprecedented move following demands from the same Chinese regulators who opposed its American listing.

DISSEMINATING AN INDUSTRY

Didi helped crystallise investors' worst fears about the brutal nature of China's campaign against its fastest-growing companies. Yet many remained in denial. How could China be trying to choke its biggest and best companies, or risk destabilising the inflow of foreign capital? As early as June 2021, fund managers at some of the largest global institutions, including Fidelity, were still declaring that China's tech clampdown was nearing an end.

'The valuations are working in favour of them because they are much less liked by other investors,' Hyomi Jie at Fidelity International said on Bloomberg TV. 'I have a view that we are closer to the end of the cycle.'

But things didn't stop with Didi.

For months, word had spread that China was looking to clamp down on another sector that had undergone rip-roaring growth in past years: its $100-billion after-school tutoring sector. Like ride-hailing and gaming, it was yet another supercharged internet-based industry raking it in by up-ending traditional business models. The rapid rise of school-curric-ulum teaching, conducted mostly via online platforms, had groomed a coterie of investment darlings that lured backing from the likes of War-burg Pincus and GIC to Temasek and Sequoia. Tech giants like Alibaba and Tencent also joined the fray.

China's online education sector was expected to generate 491 billion yuan ($76 billion) in revenue by 2024. Those lofty expectations made stock market favourites of TAL and New Oriental, and propped up giant startups like Yuanfudao and Zuoyebang.

Signs that regulators were planning to spoil the party started emerging in early 2021. My contacts in the industry had heard chatter that the government was deeply dissatisfied with the state of Chinese education – blaming the disorderly and excessive creation of tutoring institutions – and the additional workload they entailed – for exacer-bating inequity, tormenting the nation's youth, burdening parents with expensive fees, and impeding one of Xi Jinping's top priorities: reversing a declining birth rate.

The general consensus was that some form of tidying-up was in the works, but no one foresaw how serious it would become. By June, when China set up a new department to supervise the tutoring industry, people in the investment community had started hearing that the government was weighing the possibility of completely annihilating the industry.

As astonishing as that sounded, it's happened before. China's peer-to-peer finance industry was once touted as an innovative way to match savers with cash-strapped borrowers, unlocking pent-up capital and diverting it where it's needed. The sector prospered, with more than five thousand companies at its peak, attracting upwards of 50 million users and processing 3 trillion yuan of transactions a year. That industry is now in limbo, strait-jacketed by central bank officials who saw it as a ticking time-bomb.

That's why, in the following weeks, the online education industry and its many investors waited and watched with increasing apprehension. Then on a Saturday evening in July, the bomb dropped.

China ordered companies that offer tutoring on the school curriculum to go non-profit.

It was like a nuke had gone off. The State Council published a plethora of regulations that, among other things, dictated that companies that teach school subjects can no longer accept overseas investment.

Listed firms would no longer be allowed to raise capital via stock markets to invest in businesses that teach classroom subjects. Outright acquisitions were forbidden. And all vacation and weekend tutoring related to the school syllabus were now off-limits.

It sent US-listed Chinese education stocks into a tailspin, wiping 50 per cent or more off the value of sector leaders TAL Education Group, New Oriental Education & Technology Group and Gaotu Techedu Inc.

TENCENT TROUBLES

Regulators weren't done. In the second half of 2021, their gaze shifted towards the giant of social media and entertainment that Pony Ma had created.

If Alibaba and Didi were punished, there seemed to be a reasonable (or at least comprehensible) explanation – their defiance of regulators. It came as a big surprise when Tencent, a company known for being compliant and most cooperative with regulators, again stood on the receiving end of Beijing's assault.

In the second half of 2021, regulators resumed their scrutiny of the company's gaming business and slowed down the approval process of new titles once more. One state media publication went as far as to describe online games as 'spiritual opium', in a stinging rebuke of the industry.[1]

Then the country banned those under eighteen years of age from gaming more than three hours a week, further battering Tencent's shares. In response to the criticism from the government and media, Tencent curbed the time that under-18s can play, limiting them to only 14 hours of playtime during a four-week winter break in 2022.

Elsewhere, its WeChat payments business fell under inspection when regulators investigated monopolistic practices across the industry. Tencent was a big backer in many of the companies Beijing went after, including Didi, and online education platforms Yuanfudao and Huohua Siwei.

Investing in Tencent would have been one of the worst plays of 2021. Its stock plummeted 19 per cent, while the Nasdaq gained 21 per cent.

In gaming, beyond the ostensible motivation of tackling youth gaming addiction was a much more sinister issue.

The vicissitudes of Chinese political life meant Tencent could find itself in jeopardy even when following orders. One such order came from then deputy public security minister Sun Lijun, who asked Tencent to monitor some of his most powerful fellow politicians in China, according to people familiar with the matter who requested not to be named for fear of retribution.

Sun was expelled from the Communist Party in September 2021, and accused of "cultivating personal power and forming an interest group" by the country's anti-graft unit. But his surveillance endeavours emerged during the investigation, triggering a government backlash against Tencent, the people said.

The Wall Street Journal reported in February 2021 that one Tencent staff Zhang Feng was under investigation by China's antigraft watchdog starting a year prior for alleged unauthorized sharing of personal data collected by WeChat, citing people familiar. They added that Zhang was suspected of turning the data over to Sun.

---

[1] The publication later retracted the article and issued a more toned down version.

Tencent confirmed with the Journal that Zhang was under investigation, adding that the case related to "allegations of personal corruption and has no relation to WeChat or Weixin." Weixin is the local version of WeChat. Tencent also told Bloomberg at the time that the probe of Zhang did not involve Tencent's WeChat messaging service.

Zhang, who was identified as a vice president in a November 2018 statement by a local municipal government, has never held a senior position and is not a vice president, Tencent added at the time. Zhang didn't respond to a Wall Street Journal request for comment sent through China's antigraft inspectors.

Tencent declined to comment further for this book. It couldn't be determined where Zhang is now. He couldn't be reached for comment.

Another incident that put Tencent in jeopardy was that its staff launched a project using data and algorithms to predict future members of the politburo's standing committee, people familiar said. While using data to predict election outcomes is a common practice in countries like the US, it hit a nerve in China and within the Party, prompting a further probe into Tencent, the people added. One person said that the project was commissioned by Zhang and not from the top of the company, adding that it was never completed. Tencent declined to comment on the matter.

Right around that time, Pony started a long streak of no-shows for public events, between 2019 and 2020, missing key events including a high-profile artificial intelligence conference, the National People's Congress, and even the company's own annual party.

Staff internally were told that Pony was home-bound due to back injuries – the entrepreneur has had several severe disc hernias throughout his life, an issue that remains.

Unknown to the public, Tencent's experiments with data modelling and surveillance had struck a nerve within the highest echelons of the Party, because they underscored the power that Tencent possessed, according to people familiar. That meant the company had to be reined in.

Tencent had to revamp various departments across the company. In November 2021, regulators ordered the company to stop rolling out new apps or updates, declaring that its products violated data protection rules and permitting only some services to be updated after rectification.

The company is also in the process of setting up a financial holding company. Regulators are weighing on requiring Tencent to fold WeChat Payments, along with the company's insurance, lending, credit scoring and wealth management business, into a financial holding unit that will be regulated more like a bank, Bloomberg reported in March, citing people familiar. The ring fencing of WeChat Payment could get complicated, as its operations straddle two units including the messaging app and the fintech unit that provides the back-end infrastructure under the leadership of the corporate development group. The changes in set up place higher demand in capital.

Tencent has embarked on a string of divestments or asset sales, shaving interests in key elements of its empire. It handed out more than $16-billion worth of stock in China's second-largest e-commerce company, JD.com, one of its key allies. It reduced its stake in Southeast Asia's largest gaming company, the Singapore-based Sea Ltd. In addition, it reduced its stakes in a number of other US-listed Chinese stocks, including anime site Bilibili, e-commerce platform Pinduoduo and fintech company Futu Holdings Ltd.

It shuttered its in-house game streaming service in April 2022 after the Chinese regulator throttled its plans to create a giant in the space. Tencent invested in the country's two largest game streaming services and originally planned to merge them to create an equivalent of Amazon's Twitch. Tencent's in-house team would have been incorporated into that merged entity. That deal fell apart after the antitrust watchdog blocked it.

Tencent is now fully embracing what company executives have acknowledged as a "paradigm shift" in the industry. Gone are the years of reckless expansion, aggressive marketing and zero-sum competition, which partly defined the golden era of China tech. That chapter of the story has now come to an end.

Despite the turmoil, it probably was the best outcome Pony could have hoped for, as the punishment remained in the corporate realm rather than endangering Pony himself. Pony re-emerged at the annual National People's Congress meetings in 2021, proposing that the government draft laws for preserving natural reserves. A few weeks prior, he was named one of the country's most outstanding entrepreneurs.

The lesson for Pony is that he and his company have no bargaining power with the Chinese government. As authorities move to rein in its tech giants, Tencent is expected to hand over key data and at its extreme, ownership to the government, so the Party can maintain an even firmer grasp on the nation.

ATTACK ON CAPITAL

Among the notices that came out, there was a more ominous but subtle message – Xi Jinping appeared to be going after the concept of capitalism itself.

The out-of-school education industry has been 'severely hijacked by capital', according to an article posted on the website of the Ministry of Education. 'That broke the nature of education as welfare.'

Capital in China has always been somewhat of an amorphous concept. It goes beyond the institutional money often associated with large global financial firms. When such capital lands in China, it becomes interwoven with interest groups – most often backed by princelings, or the families of powerful leaders – and power factions. It's a mechanism that has lubricated the country's fastest-growing sectors from finance to consumer and tech, yet it is rarely publicised and harder to prove.

Every now and then, documents, company registries or scandals offer a glimpse into this world – the *New York Times*'s exposé about the private equity firms run by a son of former premier Wen Jiabao; the *Wall Street Journal*'s inspection of Jiang Zemin's grandson's investment vehicle Boyu Capital and its investments into Ant, Didi and Meituan; a *Financial Times* investigation into rising political star and Xi Jinping-confidante Liu He's son's Skycus and its backing of tech companies;[2] and Desmond Shum's memoir *Red Roulette: An Insider's Story*

---

[2] Eric Wen, the son of Wen Jiabao, declined to comment for that interview. His wife said reports about his business activities were unfair and inaccurate, according to the *New York Times*. Boyu didn't respond to requests for comment from the *Wall Street Journal*. Liu Tianran, Liu He's son, could not be reached for comment by the *Financial Times*.

of Wealth, Power, Corruption and Vengeance in Today's China, which provides an insider account of business and politics and the country's wealth-making machine.

Tech has been a goldmine, especially in the mobile internet era, for a plethora of the princelings operating within China's political apparatus. Venture capital firms actually make a point of hiring those luminaries to amass political clout. Many join the tech companies they invest in.

While the industry demands competence due to cutthroat competition, it's also become a breeding ground of the politically well-connected, creating a headache for regulators trying to assert control over the increasingly sprawling tech landscape.

One mid-level bureaucrat once told me that it was like poking a hornets' nest when you try to regulate China's internet companies. Everyone knows someone, every company has a powerful investor who has powerful allies even higher up. One just doesn't know who they might be offending, or worse – regulators could be standing in the way of their profits.

It's important, if not essential, that companies seek out protection, especially when they get bigger in scale. A popular saying in China is that all private companies have original sin – whether it be in the form of evading taxes, polluting the environment or bribery. That rings especially true for many internet companies that operated in sectors originally deemed illegal.

Jack Ma's Alipay operated in a grey zone during its inception, when private companies were banned from the financial sector. China's army of e-commerce platforms including Alibaba and Pinduoduo have also for years facilitated the flourishing of fake goods and counterfeit products. Ride-hailing platform Didi faced the issue of recruiting drivers and cars that don't meet standards dictated by the local governments.

In many ways, the sea-change in attitude among regulators – and mid-level bureaucrats no longer daunted by the task of cracking down on these companies – can be traced back to the humbling of Ant. Jack Ma and his crown jewel epitomised the potential landmines strewn across the tech industry. The billionaire personally commanded more political clout and global influence than any other tech mogul, and all while his company was legitimately helping place Chinese fintech on the map.

In January 2022, China's anti-corruption agency ousted the former party secretary of Hangzhou, the base of Jack Ma's Alibaba and Ant Group operations. The official was accused of serious violations of official duties, taking bribes and abuse of power.

'He colluded with capital, supported the disorderly expansion of capital,' the agency said, accusing his family members of receiving huge bribes. While not explicitly naming Jack Ma and his companies, Ant had invested in one of the companies controlled by the disgraced official's younger brother.

The signal that officials got from Ant's clampdown was that if even such a company can be messed with, then everyone else is fair game.

The change in policy is also underpinned by shifts in Xi's priorities. The assault on the tech sector mirrors crackdowns in other industries, including property. As China's economy slows and Xi tries to increase the nation's birth rate, the policies underscore the Communist Party's growing resolve to respond to mounting public dissatisfaction with hoarded wealth and narrowing avenues for advancement.

A phrase that has emerged in tandem with the crackdowns has been 'common prosperity', which refers to China's goal of becoming a modernised socialist society. After four decades of unprecedented wealth accumulation, that notion had been relegated to fable and myth, scoffed at by the ruling elite of the country.

When reformist leader Deng Xiaoping started opening up the country in the late 1970s with the notion that it was okay to 'let a few get rich first,' it was an audacious idea. It allowed China to break free from the shackles of a backward communist regime. Over the years, few have paid attention to the latter part of his speech about 'common prosperity'.

With Xi's ascendance, he's brought that back into focus.

The implications for China's tech industry are far-reaching, and could shape the playbook for the next few decades.

It could take years for the economy to fully process the effects. On the surface, the internet industry's slowdown doesn't put a dent in the real economy and manufacturing. Investment activities were still thriving in the venture capital space, reaching a record $130.6 billion in 2021, according research firm Preqin. Smaller startups continued to receive funding, as much of the regulatory curbs were focused on the internet giants. Investors turned away from platform companies and toward hardcore technologies like semiconductors, robotics and enterprise software. That aligns very much with the aspirations of the country's central planners — more breakthroughs in fundamental research and cutting edge technology that will reduce China's reliance on the U.S.

Yet on the flip side, cracks are also showing. Layoffs have ensued, with everyone from Alibaba to Tencent, and ByteDance to Didi looking for ways to tighten their belts. With the ed-tech and private-tutoring sector dismantled, one of its largest employers New Oriental dismissed 60,000 people. Then, there's that age-old question that China has to some extent continued to defy — how big of a role can China play in the knowledge economy and innovation if you restrict the flow of knowledge and information? Without the freedom of thought, how long can China's economic miracle persist?

CHAPTER 12

THE METAVERSE AND BEYOND

Meanwhile, Pony remains (relatively) active and Tencent is biding its time to shine again. One of the operations that has captured investors' imagination is its cloud business.

None of what Tencent is doing today would be possible without its cloud service, the internet infrastructure that powers WeChat and its gaming empire. Initially created to meet the internal demands of the company's billion-plus users, it's now grown into a formidable business in its own right.

As of early 2022, Tencent Cloud became the first Chinese-based company, and fifth globally, to operate more than a million servers, spreading its operations to more than seventy countries and regions.

It didn't happen overnight. Tencent executives started paving the way for the business as early as 2014, just two years after WeChat took off, at a time Amazon Web Services was only just coming into its own.

I remember sitting down with Tencent's Dowson Tong around that time, years before he would take over the company's cloud services business. Though overseeing social media services like QQ at the time, he was already clearly enthusiastic about the opportunities in the cloud, though the model inchoate.

It wasn't clear at all how the company would turn it into a viable business. Tong himself couldn't clearly articulate how they would build out the operation, what applications could run on the cloud, let alone give out any projections on how lucrative it could be. But to him it was obvious the company needed to build its own internet backbone. WeChat was undergoing exponential growth in mobile payments and community services, all of which required enormous computing power. Following that logic, it made sense to offer the same kind of capabilities to other companies. Tencent would become the first tester and user of its own cloud business.

By October that year, Tencent's ambitions in the cloud were no longer a secret. The company took to the balmy resort island of Hainan, China's Florida-equivalent, to unveil plans to set up an innovation tech park. Pony himself made the trip to the island where he grew up to headline a seminal industry gathering. There, amid the tightly packed speeches, panels and lavish pool-side parties, the company offered one of the first glimpses into how its cloud operations could become a legit business. One of the first applications the team envisioned was support for smart home appliances, connecting everything from refrigerators and lights to doors and TVs to the internet, using WeChat or QQ as the operating system. The entire affair was a turning point for not just Tencent but the industry as a whole: it signalled the geeks had not just arrived, they were on the ascendant.

It was an experiment, humble in ambition by Tencent's standards. Yet they were onto something. If you looked beyond the Fortune 500, the majority of companies in China had little idea what cloud computing was or how it could benefit them. To convince more entrepreneurs, they needed to showcase specific applications or scenarios where even smaller businesses could benefit. On top of that, working with enterprises was not Tencent's forte.

The company was best at creating platforms that catered to the masses, often building multiple similar products at the same time to see what would stick. In dealing with corporations, everything had to be streamlined. Entrepreneurs wanted solutions from one team that could represent all of Tencent.

It took some time for Tencent to find its groove. Unlike traditional cloud providers that catered to large corporations, its cloud offering's strength lay in the ability to connect small businesses with users. So if

Tencent could entice more companies to use WeChat or QQ to connect with individual users, they could get businesses to become reliant on its cloud platform. Tencent could offer them services to manage customer relations and data, human resources and conferences. These clients would in turn offer better services to their own customers, creating a beneficial feedback loop.

In other words, Tencent Cloud would grow as its customers grew. A typical example: a local shoe brand uses WeChat's Mini Programs to directly market and sell its products. It can then manage customer information, process transactions, offer promotions and dole out coupons – all via the platform.

The concept became known as the C2B2C strategy. Somewhat of an alphabet soup, C represented the end-users and daily consumers on WeChat, 2 meant 'to' and B stood for enterprises. In that sense, WeChat's teleconference and Mini App programs have been instrumental in helping the company expand its cloud business beyond just gaming companies. It's gathered enough clout to enroll the largest brands including, Nike, Gap and McDonald's.

To capture the growth in cloud, Tencent and Alibaba are once again engaged in a battle. Much of the race this time relies on hardcore technology, rather than just user interface and marketing. On the frontier, the companies are working on photonic silicon chips to combat the limitations of Moore's Law, which dictates that the number of transistors on a microchip doubles about every two years, while microprocessors grow exponentially. As the size of transistors approaches physical limits, the speed of chip development can no longer meet the data throughput demand amid the rise of high-performance computing. Bearing somewhat Asimov-novel-like qualities, silicon photonic chips promise to transfer data via light beams using photons instead of electrons. The underlying mechanic is that photons don't directly interact with each other and can travel longer distances. Another breakthrough lies within satellite-terrestrial integrated computing, essentially systems that connect satellites, aerial platforms and nautical communications networks. The appeal of this is to make digital services accessible in sparsely inhabited regions like mountainous terrain and deserts, narrowing the technology divide among the population.

BEYOND SMARTPHONES

Today, Tencent's cloud operation holds the key to another facet of its future. One of the questions its executives obsess over is what will succeed the smartphone.

While the jury is still out, Pony isn't waiting. One of the bets he's made is in automobiles. The logic is that in driverless cars, people could consume music, information, and even hold meetings as the technology matures.

That's yet to be proven, but his company acquired a 5 per cent stake in Tesla in 2017 anyway. Tencent executives thought the electric vehicle maker was undervalued back then, and its founder under-appreciated. Indeed it was. The company's stock has jumped 1,660 per cent between Tencent's purchase and March 2021. More importantly, for Tencent it was important to get its communication and entertainment apps into cars. In China, Tencent's music service, for example, is the default app on Tesla, bringing in more users and consumer data through a completely new portal.

The other more exciting possibility lies within what has come to be known as the metaverse.

When Mark Zuckerberg announced in October 2021 that Facebook was changing its name to Meta Platforms Inc., it made waves well beyond Silicon Valley. Overnight, it became the talk of the town in China as well, triggering fierce debates between founders, investors and their corporations.

The metaverse, in its simplest terms, is embodied by the world depicted in Stephen Spielberg's *Ready Player One*, where people live and play in an immersive virtual reality. It's a place where they're free to parachute off Mount Everest, ascend the Great Pyramids of Giza, race against King Kong across New York in a sports car. It's a place where people could spend the better part of their day at work, presiding over meetings with hundreds of others, even (eventually) feel another person's grip via special suits with sensors that can physically manifest their online avatars' actions in the real world.

It's no surprise that the idea of the metaverse thrilled the Chinese tech community. Every few years an overarching theme emerges, rallying talent and capital. The ability to ride such waves, or better, dictate and shape them, equates to the power to capture fortunes. The metaverse promised an entire world to explore and conquer beyond smartphones, a chance to leapfrog the giants of today that have come to dominate mobile computing.

Even on a personal level, I've witnessed over the years countless classmates and friends grow enthralled by such cycles, chasing investment bubbles in real estate and private equity, working as civil servants for the government before moving on to build startups.

Within tech, the investment themes have in the span of just a few years evolved from desktop-based social media and games, to mobile messaging to online-to-offline services, and now the metaverse. And Pony has always been a step ahead.

Pony in fact publicly laid out a vision for building something very similar to the metaverse just a few months before Zuckerberg announced his company's name change. He called it the Quan Zhen internet, meaning all-real internet. The concept, while vaguely defined, encompasses using the web to meld manufacturing and work, and overlaps with many aspects of the Facebook co-founder's vision.

'Metaverse' first appeared as a word in Neal Stephenson's 1992 novel *Snow Crash*, which depicted a world gripped by hyperinflation. The prefix 'meta' meant beyond, while 'verse' was 'universe'. It's now typically used to describe the concept of a future iteration of the internet, made up of shared three-dimensional virtual spaces.

Stephenson ironically wasn't impressed with Zuckerberg's vision, calling it 'old hat'. Crypto diehards were also quick to point out the irony of Facebook rebranding itself as Meta, as interoperability should be an inherent nature of the metaverse, meaning it should transcend a single entity.

Like most cyberpunk literature, the world that Stephenson depicted was not a kind one. Gripped by relentless inflation, people resorted to electronic payments that couldn't be taxed or traced, much like cryptocurrencies. Private organisations and entrepreneurs had taken over many of the functions of the government, including public safety and national defence. Today, we are seeing many of those predictions come to life. It's expected that cryptocurrency and the non-fungible tokens whose underlying technology could help notarise assets in the virtual world will play a big role in the metaverse. In June 2021, a patch of virtual land in the blockchain-based online world Decentraland sold for more than $900,000, hitting a record.

In *Snow Crash*, Stephenson envisioned a world based on anarcho-capitalism. And like all cyberpunk literature, the story is underlined by strong tones of anti-authoritarianism. By that token, China's metaverse could turn out very different – if allowed at all.

Assuming the government sees merit to the technology, the metaverse could in the future be split into two: China and the rest of the world. Much like the internet, China will likely shield its netizens from the rest of the global metaverse.

That China's internet industry has grown to its size today is in part because the government kept it on a loose lead while keeping it behind a firewall. The country was more preoccupied with controlling gas, oil, telecommunications, finance and traditional media. Virtually unhindered, Western capital and local entrepreneurs found opportunities in Communist China to devise a formula that married global capital and technology with the world's largest population.

But the metaverse, however, from day one, will fall under heavy scrutiny. While local government officials in cities like Shanghai seem to embrace the concept, announcing their intention to encourage its application in public services, social entertainment, games and manufacturing, others are far less sanguine. Chinese economist Ren Zeping pointed out the dangers of a metaverse, accusing it of potentially causing lower marriage and birth rates – the logic being that, if people are too busy entertaining themselves in the virtual world, they wouldn't need to seek connections in the real one.

Even matters as inherently apolitical as health policy could spur the government to shut out the metaverse. China is currently waging a battle against myopia among its youth, blaming gaming companies like Tencent for exacerbating the problem. A generation of children masked by VR headsets doesn't help the cause. And that's in addition to authorities wagging their fingers at the 'lie flat' movement, a philosophy that's gained currency among youths seeking to check out of the relentless corporate rat race.

Despite the uncertainty, companies and investors aren't holding back when it comes to investing in and preparing for what could be the next big thing. The number of applications for trademarks related to the metaverse tripled in the three months after Zuckerberg's switch, to more than eight and a half thousand in China.

Tencent's making preparations as well. Martin Lau, its president, has said that the company has the technology and know-how to build the metaverse, thanks to its enormous gaming and social media cred. It's already the Chinese local publisher for Roblox Corp.'s gaming platform, which allows users to create virtual worlds and is regarded by many as a viable early iteration of the future metaverse. Within Tencent, executives project that the industry needs at least another five years to get to the point where people will consider the technology legit. It's not that far off.

Yet there's another concept on the horizon: Web3, an internet restored to its decentralised origins by building services on top of blockchains, or managed and accessed on peer-to-peer distributed networks. The term, coined by Ethereum co-founder Gavin Wood in 2014, envisions a future where control of all data and content is put in the hands of content creators, handing back control back to the end-users, and cutting out governments and corporations. It's an all-too-familiar concept. In fact, that was the original vision of the internet when early adopters – think John Perry Barlow, the cyberlibertarian political activist – hoped they could finally be free of big institutions.

Wood argues that the web we have today is broken, corrupted from its own success. As the internet became more powerful, governments and corporations stepped in to hijack it for their own use, seizing the technology that was once meant to empower and liberate citizens. 'Society was bound to imprint itself back on the web,' Wood wrote in his 2018 blog. 'The internet today is broken by design. We see wealth, power and influence placed in the hands of the greedy, the megalomaniacs, or the plain malicious'.

The ideal version of Web3 'is an executable Magna Carta – the foundation of the freedom of the individual against the arbitrary authority of the despot'.

The promises of Web3 are enticing enough, so much so that corporations and venture capitalists are already rushing into the space, creating memes that have made the concept somewhat amorphous. Sceptics argue it's ironically perverting the notion of Web3, as they're betting that it will yield the next Apple Inc. – reasserting corporate dominance in a realm that's supposed to champion the little guy. Former Twitter Chief Executive Officer Jack Dorsey has become one of the biggest critics, arguing that far from democratising the web, this current frenzy is but another tool of venture capitalists. Elon Musk has said it's just marketing hype.

Corporations from the Web2 era are dispatching teams internally to develop their own versions of Web3, incorporating such services onto their own platform to stay relevant.

Tencent is no exception in that it's brainstorming ways to stake out a plot in the future internet landscape, but it faces unique constraints.

Developing Web3 in China is almost impossible, as everything about the concept contradicts the government's central priorities, namely maintaining control over content and infrastructure. And by limiting its own technology corporations, China might lose out on what could become the next wave.

That has to be enormously frustrating for someone like Pony.

The self-proclaimed geek who, as a child, gazed at the stars and pondered how he could better the universe, has in a sense realised his goal and finds himself now at a crossroads, dwelling on what legacy he will leave behind. He found his calling as one of the pioneers of the global mobile internet, linking billions to vast new realms of on-the-go mobile entertainment and communications. He helped solidify emotional connections between lovers, families and friends even as they ventured across the Pacific, helping them keep in touch and see each other.

When my father left the country in the 1980s to pursue his studies overseas, his only means of staying in touch with my mother was one letter every few months. To them, it felt like once a person left the country, it could mean losing touch forever. Technology has changed everything. During the pandemic, WeChat became the primary means for loved ones to keep in touch with each other on a daily, if not hourly basis. Tencent's products in gaming have soothed many lonely souls, inspired the imagination, and even helped others find camaraderie in their virtual quests together. And life has become so convenient that the only device one needs when they travel in China is a phone. In the process, Pony gained unimaginable wealth and influence beyond his wildest dreams.

Now everything may be out of his hands. The empire he's created may have grown too vast and powerful for the Chinese government to stomach, a beast that needs to be tamed, a tool that needs to be harnessed to ensure the rule of the Party. In that environment, attempting to blaze new trails could be a minefield.

The milieus of *Ready Player One* and *Snow Crash* already don't inspire the most faith in humanity's future, depicted as a soul-crushing struggle against all-seeing, all-powerful institutions. It's a vision that some would argue is materialising in China, which has put in place the world's largest and most effective surveillance and control mechanisms.

Pony's conundrum is how to propel Tencent into the future while appeasing his political masters – an incredibly delicate manoeuvre with unimaginable stakes. Given the extraordinary achievements of the past two decades, some may argue Pony is duty-bound to try. Who better than the visionary founder of the world's largest online entertainment empire to square that circle, to fashion a formula that will work for a fifth of the global population?

And so the billionaire might not be bowing out anytime soon. Perhaps it's not even up to him. For the Party, it would be much easier to have one mighty all-knowing yet obedient company to rule all than play whack-a-mole with potentially disruptive forces.

Some choices Pony's made following China Tech Inc.'s annus horribilis offer clues to his thinking for the next decade, or at least an overarching stance for public consumption and his political overlords.

Back in 2010, when Pony made the crucial decision to reorganise Tencent and open up the company's platform, he declared that he wanted the company to become an infrastructure-like company, to become the equivalent of water and electricity for the internet.

Fast-forward a decade, Pony has made a U-turn, telling staff at an internal 2021 year-end gathering that Tencent is just a dime a dozen, a beneficiary of the vast progress the country's made.

'Tencent is not an infrastructure-service company and can be replaced at any given moment,' local media Late Post reported him saying. 'In the future, when Tencent services the country and society', the company needs to make sure it 'doesn't overstep, be a good assistant'.

In the *I Ching*, also known as the *Book of Changes*, the ancient Chinese script confers a note of wisdom: that an overconfident dragon will have cause for remorse. If Pony's generation of entrepreneurs spent the first half of their lives swimming upstream like carps to leap over the gates of the Yellow River hoping to become dragons, then, following ancient wisdom, the second part of their fate lies in knowing when to bow out – or perhaps becoming part of the system they once wished to change.

ACKNOWLEDGEMENTS

This book would not have been possible without the people who supported and mentored me through life.

I first and foremost would like to thank Brad Stone. Without him, I didn't even know I had it in me to write a book. Brad, author of Amazon Unbound and senior executive editor of Bloomberg's technology team, first proposed the idea of me writing this book when we were reporting a Businessweek story on Tencent in June 2017. Despite his busy schedule of running a global team, he always found time to talk to me when I needed guidance. Thank you Brad for pushing me to pursue bigger and more ambitious stories, and making me a better reporter.

Huw Armstrong and Hodder & Stoughton, Hachette made this project a reality. Huw championed this book, noticing the importance of Tencent and China Tech Inc. before others. My agent Marysia Juszczak-iewicz has been a relentless force in securing a publisher and keeping the project on track. Her company Peony Literary Agency has represented some of the most important books covering the subject of China.

I'm eternally grateful to my teachers and editors who have had an important impact on me. Madam Hao Youming for imparting me with the right values; Chris Hawke, Howard French, Mirta Ojito, Jennifer Preston for igniting the fire of journalism within me, and showing me the greatness and importance of our craft. David Lague, SK Witcher,

Kenneth Howe, Gina Chua for taking a leap of faith in me when I was still a rookie, and helping me build the foundation to become a better reporter.

It's been a blast battling in the trenches to cover China tech with Peter Elstrom, Michael Tighe, Rob Fenner, David Ramli and Zheping Huang. I couldn't have hoped for a more brilliant, caring and dedicated team of colleagues.

I also want to thank my editors at Bloomberg who've advocated for and mentored me: Candice Zacharias, Michael Patterson, Tom Giles, David Scanlan, David Scheer, Caroline Gage, Hwee Ann Tan, Anand Krishnamoorthy, John Liu, Heather Harris, Sarah Wells and Madeleine Lim.

Thank you everyone at Tencent and the Chinese tech industry who took time out of your busy schedules to sit down with me or answer my fact checking questions. I have tremendous respect for reporters, especially local ones, who have covered the beat with dedication. Their works have been cited in the endnotes.

If you've seen buzz about this book, it's because of Hodder and Stoughton's publicist Ollie Martin, marketing expert Sahina Bibi and Bloomberg's Rob Koh. Nick Fawcett copy edited and Helena Caldon proofread the manuscript.

I am fortunate to have such loving parents who supported me fully. You've worked hard all your lives so I could follow my heart to pursue my passion.

And most of all I want to thank my husband. My friend joked that I was having two babies, working on this book while pregnant. And indeed, the process of writing it was like giving life to something. It was a memorable experience to conduct interviews all the way up to the same day my daughter was born and jumping back to action within the same week. My husband encouraged me when I felt like giving up, calmed my nerves, acted as my sounding board, first reader and editor. I could not have done any of this without his support.

This book is dedicated to him and our daughter.

NOTES

作者:《腾讯十年》创作组. 企鹅传奇. 出版社: 深圳报业集团出版社

作者: 熊江. 小QQ大帝国: 马化腾传奇. 出版社: 中央编译出版社

失控的腾讯帝国. 时代周报. https://www.163.com/tech/article/7LAHRGJ000094LN8. html

魏武挥. 诊断腾讯: 眼中没有公众难成伟大企业. 第一财经日报.

2021, D. W. T. (2022, January 13). *The Games Market and Beyond in 2021: The Year in Numbers.* Newzoo. https://newzoo.com/insights/articles/the-games-market-in-2021-the-year-in-numbers-esports-cloud-gaming/

Barboza, D. (2012, October 26). *Family of Wen Jiabao Holds a Hidden Fortune in China.* The New York Times. https://www.nytimes.com/2012/10/26/business/global/family-of-wen-jiabao-holds-a-hidden-fortune-in-china.html

Billionaire Donations Soar in China Push for 'Common Prosperity.' (2021). Bloomberg. https://www.bloomberg.com/tosv2.html?vid=&uuid=82d5f3c6-d45a-11ec-b92a-5772676d6d73&url=L25ld3MvYXJ0aWNsZXMvMjAyMS0wOC0yNi9iaWWx-saW9uYWlyZS1kb25hdGlvbnMtc29hci1pbi1jaGluYS1wdXNoLWZvci1jb21t-b24tcHJvc3Blcml0eT9zcmVmPWhtbzzdncWl5

BloombergBillionairesIndex.(n.d.).Bloomberg.https://www.bloomberg.com/tosv2.html?vid=&uuid=eec88849-d453-11ec-bcb7-6b6769587048&url=L2JpbGxpb25haXJlcy9wcm9maWxlcy90aW0tc3dlZyeS8/c3JlZj1obW83Z3FpeQ==

Bloomberg News. (2020, November 6). *Inside the Chaotic Unraveling of Jack Ma's $35 Billion IPO.* BQ Prime. https://www.bqprime.com/china/inside-the-chaotic-unraveling-of-jack-ma-s-35-billion-ant-ipo

Bloomberg News. (2021, July 23). *China Weighs Unprecedented Penalty for Didi After U.S. IPO.* BQ Prime. https://www.bqprime.com/business/china-is-said-to-weigh-unprece-dented-penalty-for-didi-after-ipo

Bloomberg News. (2022, January 27). *China's Communists Oust First Official Over 'Disorderly' Capital.* BQ Prime. https://www.bqprime.com/global-economics/commu-nist-party-expels-hangzhou-chief-over-disorderly-capital

Blumberg, P. (2020). *WeChat Judge Won't Pause Temporary Order Blocking Trump Ban.* Bloomberg. https://www.bloomberg.com/news/articles/2020-10-23/wechat-judge-won-t-pause-temporary-order-blocking-trump-ban?sref=hmo7gqiy

Boren, C. (2019). *shaquille-oneal-weighs-nba-china-controversy-daryl-morey-was-right.* Washington Post. https://www.washingtonpost.com/sports/2019/10/23/shaquille-oneal-weighs-nba-china-controversy-daryl-morey-was-right/

C, L. (2013). *Tencent Buys $448 Million Stake in Sohu Unit to Win Users.* Bloomberg. https://www.bloomberg.com/tosv2.html?vid=&uuid=13864553-d451-11ec-8d84-644e75454547&url=L25ld3MvYXJ0aWNsZXMvMjAxMy0wOS0xNi90ZW5jZW50LWJ1eXMtNDQ4L-W1pbGxpb24tc3Rha2UtaW4tc29odS1zZWFyY2gtdW5pdC10by13aW4tdXNlcnM/c3JlZj1obW83Z3FpeQ==

C, L. (2019a). *Alibaba Seals $38 Billion Singles' Day Sales Record.* Https://Www.Bloomberg.Com/News/Articles/2019-11-10/Alibaba-Hits-6-Billion-Yuan-of-Singles-Day-Sales-in-a-Minute?Sref=hmo7gqiy. https://www.bloomberg.com/tosv2.html?vid=&uuid=0e54a6bf-d42f-11ec-8b21-776270434b51&url=L25ld3MvYXJ0aWNsZXMvMjAxOS0xMS0xMC9hbGliYWJhLWhpdHMtNi1iaWxsaW9uLXll1YW4tb2Ytc2luZ2xlcy1kYXktc2FsZXMtaW4tYS1taW51dGU/c3JlZj1obW83Z3FpeQ==

C, L. (2019b). *The World's Greatest Delivery Empire.* Businessweek. https://www.bloomberg.com/tosv2.html?vid=&uuid=6069b05b-d430-11ec-8703-4b6958736b74&url=L2ZlYXR1cmVzLzIwMTktbWVpdHVhbi1jaGluYS1zkZWxpdmVyeS1lbXBpcmUvP3NyZWY9aG1vN2dxaXk=

失控的腾讯帝国.   2015.时代周报.   https://www.163.com/tech/article/7LAHRG-J000094LN8.html

Chen, K. (2003a, October 31). *Chinese Wedding Pulls Back Veil Of Secrecy on Money-Power Unions.* WSJ. https://www.wsj.com/articles/SB106754823050625000

Chen, K. (2003b, October 31). *Chinese Wedding Pulls Back Veil Of Secrecy on Money-Power Unions.* WSJ. https://www.wsj.com/articles/SB106754823050625000

Chen, L. (2016). *Ninja Naruto Leads Tencent's March into China's $31 Billion Anime Market.* Bloomberg. https://www.bloomberg.com/tosv2.html?vid=&uuid=e519d706-d457-11ec-bbc6-424f47775358&url=L25ld3MvYXJ0aWNsZXMvMjAxNi0wMy0xNS9uaW5qYS1uYXJ1dG8tbGVhZHMtdGVuY2VudC1zLW1hcmNoLWludG8tY2hpbmEtcy0zMS1iaWxsaW9uLWFuaW1lLW1hcmtldD9zcmVmP-WhtbzdncWl5

Chen, L. (2017). *Tencent Music Drowns Out Spotify and Apple in China.* Bloomberg. https://www.bloomberg.com/tosv2.html?vid=&uuid=94bf1df8-d455-11ec-ab58-7951707a5267&url=L25ld3MvYXJ0aWNsZXMvMjAxNy0xMi0xMy90ZW5jZW50LWZlbnRRW9mZ90aW5nLW9kZ90aW5nLW9kZmcwcml0LWluLWNoaW5hLXMtYm0dGxlLWZvci10aGUtYmFzZHM/c3JlZj1obW83Z3FpeQ==

Chen, L. (2018a). *China Freezes Game Approvals Amid Agency Shakeup.* Bloomberg. https://www.bloomberg.com/news/articles/2018-08-15/china-is-said-to-freeze-game-approvals-amid-agency-shakeup

Chen, L. (2018b). *Tencent Slashes Game Marketing Budget Amid Freeze.* Bloomberg. https://www.bloomberg.com/tosv2.html?vid=&uuid=5a212ffd-d458-11ec-ba60-4a626d49615

5&url=L25ld3MvYXJ0aWNsZXMvMjAxOC0xMS0wOC90ZW5jZW50LWlzLX-
NhaWQtdG8tc2xhc2gtZ2FtZS1tYXJrZXRpbmctYnVkZ2V0LWFtaWQtZnJlZX-
plP3NyZWY9aG1vN2dxaYk=

Chen, L. (2019). *Tencent Gets 'Wakeup Call' From China's Assertions of Patriotism*. Bloomb-
    erg. https://www.bloomberg.com/tosv2.html?vid=&uuid=f2f1d2fd-d458-11ec-adeb
    -615766545966&url=L25ld3MvYXJ0aWNsZXMvMjAxOS0xMC0xMS90ZW5j-
    ZW50LWdldHMtd2FrZXVwLWNhbGwtZnJvbS1jaGluaYS1zLWFzc2VydGlvbn-
    Mtb2YtcGF0cmlvdGlzbT9zcmVmPWhtbzdncWl5

Chen, L. (2021). *China's EdTech Assault Hits Investors From Tiger to Temasek*. Bloomberg.
    https://www.bloomberg.com/news/articles/2021-07-23/from-tiger-to-temasek-inves-
    tors-scarred-by-china-edtech-assault?sref=hmo7gqiy

Chen, L., Alpeyev, P., & Nakamura, Y. (2016). *China's Tencent Buys Control of Clash of Clans Maker
    for $8.6 Billion*. Bloomberg. https://www.bloomberg.com/news/articles/2016-06-21/ten-
    cent-leads-8-6-billion-deal-for-clash-of-clans-game-studio?sref=hmo7gqiy

Chen, L., & Liu, C. (2020). *How China Lost Patience With Jack Ma, Its Loudest Billionaire*.
    Businessweek. https://www.bloomberg.com/tosv2.html?vid=&uuid=2bf1f0cd-d45a-
    11ec-89a3-6e7554546161&url=L25ld3MvZmVhdHVyZXMvMjAyMC0xMi0yM-
    i9qYWNrLW1hLXMtZW1waXJlLWluLWNyaXNpcy1hZnRlci1jaGluaYS1oY-
    Wx0cy1hbnQtZ3JvdXAtaXBvP3NyZWY9aG1vN2dxaYk=

Cheng, E. (2021, December 31). *Shanghai doubles down on the metaverse by including it in a
    development plan*. CNBC. https://www.cnbc.com/2021/12/31/shanghai-releases-five-
    year-plans-for-metaverse-development.html

China Fines Alibaba Record $2.8 Billion After Monopoly Probe. (2021). Bloomberg. https://
    www.bloomberg.com/tosv2.html?vid=&uuid=6175dd74-d45a-11ec-b5a6-71494b414
    34d&url=L25ld3MvYXJ0aWNsZXMvMjAyMS0wNC0xMC9jaGluYS1maW5lcy-
    1hbGliYWJhLWdyb3VwLTItOC1iaWxsaW9uLWluLW1vbm9wb2x5LXByb2Jl

China Just Became the Games Industry Capital of the World. (2017). Bloomberg. https://www.
    bloomberg.com/tosv2.html?vid=&uuid=72931fdb-d450-11ec-b72b-4265774b5571&
    url=L25ld3MvYXJ0aWNsZXMvMjAxNy0wNi0wMS9jaGluYS1qdXN0LWJlY-
    2FtZS10aGUtZ2FtZXMtaW5kdXN0cnktY2FwaXRhbC1vZi10aGUtd29ybGQ/
    c3JlZj1obW83Z3FpeQ==

China Removes Key Hurdle to Allow U.S. Full Access to Audits. (2022). Bloomberg.
    https://www.bloomberg.com/tosv2.html?vid=&uuid=8d385815-d459-11ec-
    b430-664375674c64&url=L25ld3MvYXJ0aWNsZXMvMjAyMi0wNC0w
    Mi9jaGluYS1yZW1vdmVzLW1ham9yLWh1cmRsZS10by1hbGxvdy11LXMtZnVs-
    bC1hY2Nlc3MtdG8tYXVkaXRzP3NyZWY9aG1vN2dxaYk=

China Weighs Tencent Payments Overhaul, New License Requirement. (2022). Bloomberg. https://
    www.bloomberg.com/tosv2.html?vid=&uuid=44900d6c-d45c-11ec-b0c6-4e61757777
    55&url=L25ld3MvYXJ0aWNsZXMvMjAyMi0wMy0xOC9jaGluYS13ZWlnaHMt
    dGVuY2VudC1wYXltZW50cy1vdmVyaGF1bC1uZXctbGljZW5zZS1yZXF1aXJl-
    WVudD9zcm5kPXByZW1pdW0tYXNpYSZzcmVmPWhtbzdncWl5

China's Didi Crackdown Is All About Controlling Big Data. (2021). Bloomberg. https://www.
    bloomberg.com/tosv2.html?vid=&uuid=6c6ce1e8-d45b-11ec-a882-42444e4d5164

&url=L25ld3MvZmVhdHVyZXMvMjAyMS0wNy0wOC8tZGlkaS1jcmFja2Rvd-24tYmlnLWRhdGEtaXMtdGhlLWxhdGVzdC11LXMtY2hpbmEtYmF0dGxlZ-3JvdW5kP3NyZWY9aG1vN2dxaXk=

Chinese pair make breakthrough in Team Championship. (2018). Alfred Dunhill Links Championship. https://www.alfreddunhilllinks.com/news/chinese-pair-make-break-through-in-team-championship/

Crecente, B. (2019). League of Legends is now 10 years old. This is the story of its birth. Washington Post. https://www.washingtonpost.com/video-games/2019/10/27/league-leg-ends-is-now-years-old-this-is-story-its-birth/

Desk, T. (2021, May 6). Tim Sweeney: 11 facts you didn't know about the CEO of Epic Games. The Indian Express. https://indianexpress.com/article/technology/tech-news-technology/tim-sweeney-interesting-facts-you-didnt-know-about-the-ceo-of-epic-games-7304178/

The Epic Games store is now live. (n.d.). Epic Games. https://store.epicgames.com/en-US/news/the-epic-games-store-is-now-live

Fogel, S. (2019). Epic's Support-A-Creator Program Expanding to All Titles on Its Store. Variety. https://variety.com/2019/gaming/news/epic-games-support-a-creator-program-ex-panding-expands-1203129403/

Fortnite exits China over crackdown. (2021, November 4). Young Post. https://www.scmp.com/yp/discover/entertainment/tech-gaming/article/3154829/fortnite-exits-chi-na-over-crackdown

Fortnite Fight: CEO Explains Why He Launched War Against Apple, Google. (2020, September 10). NPR. https://choice.npr.org/index.html?origin=https://www.npr.org/2020/09/10/911658041/fortnite-maker-tim-sweeney-on-apple-and-google-these-monopolies-need-to-be-stopp

Frank, A. (2015, December 17). Riot Games now owned entirely by Tencent. Polygon. https://www.polygon.com/2015/12/16/10326320/riot-games-now-owned-entirely-by-tencent

Fung, B. (2019, January 18). Netflix: Fortnite is a bigger rival than HBO. Washington Post. https://www.washingtonpost.com/technology/2019/01/18/netflix-fortnite-is-bigger-rival-than-hbo/

The future is Unreal (Engine) | Their future is Epic: The evolution of a gaming giant. (n.d.). Polygon.Com. https://www.polygon.com/a/epic-4-0/the-future-is-unreal-engine

Gamers Boycott Blizzard After Protest Sympathizer Is Banned. (2019). Bloomberg. https://www.bloomberg.com/tosv2.html?vid=&uuid=d8e1cfab-d458-11ec-bcb7-6b6769587048&url=L25ld3MvYXJ0aWNsZXMvMjAxOS0xMC0wOS9nYW1lcn-MtY2FsbC1mb3ItYm95Y290dC1vZi1ibGl6emFyZC1hZnRlci1ob25nLWtvb-mctcHJvdGVzdC1iYW4/c3JlZj1obW83Z3FpeQ==

Handrahan, M. (2019, January 18). Netflix: "We compete with (and lose to) Fortnite more than HBO." GamesIndustry.Biz. https://www.gamesindustry.biz/articles/2019-01-18-netf-lix-we-compete-with-and-lose-to-fortnite-more-than-hbo

Hong, J., Chen, L., & G, Y. (2019). Tencent Airs NBA Games as Chinese State TV Blackout Persists. Bloomberg. https://www.bloomberg.com/tosv2.html?vid=&uuid=b156a03e-d458-11ec-aebc-744c6f6f646a&url=L25ld3MvYXJ0aWNsZXMvMjAxOS0xMC0x-NC90ZW5jZW50LWFpcnMtdHdvLW5iYS1nYW1lcy1hcy1jaGluZXNlLXN0Y-XRlLXR2LWJsYWNrb3V0LXBlcnNpc3RzP3NyZWY9aG1vN2dxaXk=

Howcroft, E. (2021, June 18). *Virtual real estate plot sells for close to $1 mln.* Reuters. https://www.reuters.com/technology/virtual-real-estate-plot-sells-close-1-mln-2021-06-18/

Huang, Z. (2021). *Tencent Employee Faces Corruption Investigation in China.* Bloomberg. https://www.bloomberg.com/news/articles/2021-02-11/tencent-executive-held-by-china-over-corruption-probe-wsj-says?sref=hmo7gqiy

Hytha, M. (2018). *Spotify Gets a Boost as Tencent Music Kicks Off Share Sale.* Bloomberg. https://www.bloomberg.com/tosv2.html?vid=&uuid=b410773c-d455-11ec-b72b-4265774b5571&url=L25ld3MvYXJ0aWNsZXMvMjAxOC0xMi0wNC9zcG90aWZ5LWdldHMtYS1ib29zdC1hcy10ZW5jZW50LW11c2ljLWtpY2tzLW9mZi1zaGFyZS1zYWxlP3NyZWY9aG1vN2dxaXk=

Inside Tencent's Gambit to Dominate a $13 Billion Esports Arena. (2018). Bloomberg. https://www.bloomberg.com/tosv2.html?vid=&uuid=bedf24e5-d457-11ec-8703-4b6958736b74&url=L25ld3MvYXJ0aWNsZXMvMjAxOC0wNy0yNC9pbnNpZGUtdGVuY2VudC1zLWdhbWJpdC10by1kb21pbmF0ZS1hLTEzLWJpbGxpb24tZXNwb3J0cy1hcmVuYT9zcmVmPWhtb2dzncWl5

Isaac, M. (2017, March 3). *How Uber Deceives the Authorities Worldwide.* The New York Times. https://www.nytimes.com/2017/03/03/technology/uber-greyball-program-evade-authorities.html

Kharif, O. (2020, August 26). *Fortnite's Leader Makes a Career Crusading Against Big Tech.* Bloomberg. https://www.bqprime.com/technology/fortnite-s-sweeney-makes-career-out-of-crusade-against-big-tech

Knight, S. R. A. (2013, October 20). *SoftBank buys $1.5 billion stake in Finnish mobile games maker Supercell.* U.S. https://www.reuters.com/article/net-us-softbank-acquisition-idINBRE99E0ID20131020

Li, P. B. G. (2019, May 28). *After "Honour of Kings" failure abroad, Tencent retools overseas strategy.* U.S. https://www.reuters.com/article/us-tencent-holdings-videogames/after-honour-of-kings-failure-abroad-tencent-retools-overseas-strategy-idUKKCN1SX1Y0?edition-redirect=uk

The life and rise of Tim Sweeney, the billionaire CEO behind "Fortnite" who's now taking on Apple in a lawsuit that could have huge implications for the whole industry. (2020, August 19). Business Insider Nederland. https://www.businessinsider.nl/fortnite-maker-epic-games-ceo-tim-sweeney-history-timeline-2019-10?international=true&r=US

Liu, C. (2021). *China Widens Internet Crackdown With Meituan Monopoly Probe.* Bloomberg. https://www.bloomberg.com/news/articles/2021-04-26/china-investigates-meituan-for-suspected-monopolistic-practices?sref=hmo7gqiy

Liu, C. (2022, January 11). *China Venture Funding Hits Record $131 Billion Despite Crackdown.* BQ Prime. https://www.bqprime.com/china/china-venture-funding-hits-record-131-billion-despite-crackdown

Ma, W., & Osawa, J. (2020, December 21). *Tensions Flare Behind the Scenes of 'League of Legends.'* The Information. https://www.theinformation.com/articles/tensions-flare-behind-the-scenes-of-league-of-legends?rc=3jagi4://lol.garena.com/news/articles/782

A mysterious message millionaire. (2009). China Daily. http://www.chinadaily.com.cn/business/2009-01/12/content_7388202.htm

Nakamura, Y., & Kim, S. (2017). *One Man's Journey From Welfare to World's Hottest Video Game.* Bloomberg. https://www.bloomberg.com/news/articles/2017-09-27/one-man-s-journey-from-welfare-to-world-s-hottest-video-game?sref=hmo7gqiy

Nayak, M. (2021). *Epic CEO Denies Attack on Apple App Store Is to Boost Fortnite.* Bloomberg. https://www.bloomberg.com/news/articles/2021-05-03/epic-ceo-says-lifting-fortnite-sales-isn-t-why-he-s-suing-apple?sref=hmo7gqiy

No Chicken? Tencent's PUBG Stand-In Leaves Gamers Fuming. (2019). Bloomberg. https://www.bloomberg.com/news/articles/2019-05-15/no-chicken-tencent-s-pubg-stand-in-leaves-gamers-fuming?sref=hmo7gqiy

NPR Cookie Consent and Choices. (2020, December 10). NPR. https://choice.npr.org/index.html?origin=https://www.npr.org/2020/12/10/944978668/riot-games-brandon-beck-and-marc-merrill

One Driver Explains How He Is Helping to Rip Off Uber in China. (2015). Bloomberg. https://www.bloomberg.com/tosv2.html?vid=&uuid=c06109bf-d44f-11ec-8879-62784f5a6743&url=L25ld3MvYXJ0aWNsZXMvMjAxNS0wNi0yOC9vbmUtZHJpdmVyLWV4cGxhaW5zLWhvdy1oZS1pcy1oZWxwaW5nLXRvLXJpcC1vZmYtdWJlci1pbi1jaGluYT9zcmVmPWhtbzdncWl5

Palmeri, C. (2018). *Millions of Women Will Make Fortnite a Billion-Dollar Game.* Bloomberg. https://www.bloomberg.com/tosv2.html?vid=&uuid=ea4049ba-d454-11ec-aa60-66775079677a&url=L25ld3MvYXJ0aWNsZXMvMjAxOC0wNS0yOS9taWxsaW9ucy1vZi13b21lbi13aWxsLW1ha2UtZm9ydG5pdGUtYS1iaWxsaW9uLWRvbGxhcGci1nYW1lP3NyZWY9aG1vN2dxaXk=

Palumbo, A. (2019, February 2). *Epic's Tim Sweeney Defends Epic Games Store and Himself, Says Exclusives Are a Legitimate Way to Compete.* Wccftech. https://wccftech.com/tim-sweeney-defends-epic-games-store/

Ping, C. K., & Yu, X. (2020, December 9). *China Hails Victory in Crackdown on Peer-to-Peer Lending.* WSJ. https://www.wsj.com/articles/china-hails-victory-in-crackdown-on-peer-to-peer-lending-11607515547

Rubio Requests Information from MSCI Over Controversial Decision to add Chinese Companies in its Equity Indexes. (2019). U.S. Senator for Florida, Marco Rubio. https://www.rubio.senate.gov/public/index.cfm/2019/6/rubio-requests-information-from-msci-over-controversial-decision-to-add-chinese-companies-in-its-equity-indexes

Savov, V., & Huang, Z. (2021). *League of Legends and Twitch Streamers Fuel Latest Netflix Hit.* Bloomberg. https://www.bloomberg.com/tosv2.html?vid=&uuid=22ec4241-d457-11ec-837f-7458656b6455&url=L25ld3MvYXJ0aWNsZXMvMjAyMS0xMS0wOC90ZW5jZW50LWdldHMtcVzcGl0ZS1mcm9tLWNyYWN-rZG93bi13aXRoLWxlYWd1ZS1vZi1sZWdlbmRzLWhpdHMvc3JlZjlobW83Z3F-peQ==

Searching for the Next Jack Ma. (2015). Markets Magazine. https://www.bloomberg.com/tosv2.html?vid=&uuid=fc7a8e6f-d44f-11ec-86ed-55614b776a63&url=L25ld3M-vYXJ0aWNsZXMvMjAxNS0wNS0xMS9jaGluZXNlLWludmVzdG1lbnQtYm-Fua2VyLWZhbiliYW8tc2VhcmNoZXMtZm9yLW5leHQtamF5ltYQ==

Staff, W. S. J. (2021, February 11). *Tencent Executive Held by China Over Links to Corruption Case*. WSJ. https://www.wsj.com/articles/tencent-executive-held-by-china-over-links-to-corruption-case-11613009016

Staff, X. W. (2015, October 20). *Microsoft Studios acquires rights to Gears of War franchise*. Xbox Wire. https://news.xbox.com/en-us/2014/01/27/games-microsoft-studios-gears-of-war/

Statista. (2021, August 12). *Global box office revenue coronavirus impact 2020–2025*. https://www.statista.com/statistics/1170721/impact-coronavirus-global-box-office-revenue/

Statista. (2022, February 21). *Fortnite player count 2020*. https://www.statista.com/statistics/746230/fortnite-players/

Statt, N. (2020, May 6). *Fortnite is now one of the biggest games ever with 350 million players*. The Verge. https://www.theverge.com/2020/5/6/21249497/fortnite-350-million-registered-players-hours-played-april

Supercell. (n.d.). Supercell. https://supercell.com/en/about-us/

Tencent Closes Game Streaming Site After Beijing Blocks Merger. (2022). Bloomberg. https://www.bloomberg.com/news/articles/2022-04-07/tencent-closes-game-streaming-site-after-beijing-blocks-merger?sref=hmo7gqiy

Tencent Cloud: Creating Global Prosperity. (n.d.). Bloomberg. https://www.bloomberg.com/tosv2.html?vid=&uuid=b14cbbbf-d45a-11ec-88ae-57796a584656&url=L-2FydGljbGUvdGVuY2VudC1jbG91ZC90ZW5jZW50LWNsb3VkLWNyZW-F0aW5nLWdsb2JhbC1wcm9zcGVyaXR5P3V0bV9tZWRpdW09c29jaWFsJnV0b-V9pZD1jdXN0b21jb250ZW50LQ==

Tencent, JD.com Said in Talks to Combine E-Commerce Business. (2014). Bloomberg. https://www.bloomberg.com/tosv2.html?vid=&uuid=9a21dda0-d451-11ec-8e4f-677063625045&url=L25ld3MvYXJ0aWNsZXMvMjAxNC0wMi-0yMC90ZW5jZW50LWpkLWNvbS1zYWlkLWluLXRhbGtzLXRvLWN-vbWJpbmUtZS1jb21tZXJjZS1idXNpbmVzcy0xLT9zcmVmPWhtbzdncWl5

Tim Sweeney - Top podcast episodes. (n.d.). Listen Notes. https://www.listennotes.com/top-podcasts/tim-sweeney/

Tim Sweeney Answers Questions About The New Epic Games Store. (2018, December 4). Game Informer. https://www.gameinformer.com/2018/12/04/tim-sweeney-answers-questions-about-the-new-epic-games-store

Tshabalala, S. (2015a, July 27). *Africa's biggest media group, Naspers, has finally apologized for its role during Apartheid*. Quartz. https://qz.com/africa/464343/africas-biggest-media-group-naspers-has-finally-apologized-for-its-role-during-apartheid/

Tshabalala, S. (2015b, July 27). *Africa's biggest media group, Naspers, has finally apologized for its role during Apartheid*. Quartz. https://qz.com/africa/464343/africas-biggest-media-group-naspers-has-finally-apologized-for-its-role-during-apartheid/

Uber Slayer: How China's Didi Beat the Ride-Hailing Superpower. (2016). Businessweek. https://www.bloomberg.com/tosv2.html?vid=&uuid=9c779c2f-d44f-11ec-a113-47646563744e&url=L2ZlYXR1cmVzLzIwMTYtZGlkaS1jaGVuZy13ZWkvP3Ny-ZWY9aG1vN2dxaXk=

Ubisoft to Release Tom Clancy's The Division 2 on Epic Games Store. (n.d.). Epic Games. https://store.epicgames.com/en-US/news/ubisoft-to-release-tom-clancy-s-the-division-2-on-epic-games-store

U.S. expels WeChat, TikTok from app stores on China concern - BNN Bloomberg. (2020, September 18). BNN. https://www.bnnbloomberg.ca/u-s-expels-wechat-tiktok-from-app-stores-on-china-concern-1.1495945

U.S. Teachers and Firefighters Are Funding Rise of China Tech Firms. (2019). Bloomberg. https://www.bloomberg.com/tosv2.html?vid=&uuid=94cd9e8c-d458-11ec-8f83-6969666a4755&url=L2dyYXBoaWNzLzIwMTktY2hpbmEtdGVjaC1tb25leS10aHJlYXRlbmVkLWJ5LXRyYWRlLXdhci8=

van Boom, D. (2015, October 15). *Korean video game Crossfire brought to the big screen by "Fast & Furious" makers.* CNET. https://www.cnet.com/tech/gaming/south-korean-fps-crossfire-to-get-film-adaptation-by-fast-furious-makers/

Vynck, G. (2020). *WeChat iPhone Downloads Surge in the U.S. Ahead of Trump Ban.* Bloomberg. https://www.bloomberg.com/tosv2.html?vid=&uuid=ab5de96e-d459-11ec-9a37-536547784766&url=L25ld3MvYXJ0aWNsZXMvMjAyMC0wOS0xOC93ZWNoYXQtaXBob25lLWRvd25sb2Fkcy1zdXJnZS1pbi10aGUtdS1zLWFoZWFkLW9mLXRydW1wLWJhbi9zcmVmPWhtbWdncW4cWl5

Wood, G. (2018, September 20). *Why We Need Web 3.0 - Gavin Wood.* Medium. https://gavofyork.medium.com/why-we-need-web-3-0-5da4f2bf95ab

Wu, A., & Zhang, C. (2019). *Epic Games - Case - Faculty & Research - Harvard Business School.* Harvard Business School. https://www.hbs.edu/faculty/Pages/item.aspx?num=55889

Wu, X. (2007). *Chinese Cyber Nationalism: Evolution, Characteristics, and Implications.* Lexington Books.

Ye, J. (2021, September 10). *China's regulators said to slow their approval of new online games, as Beijing's campaign against gaming. . .* South China Morning Post. https://www.scmp.com/tech/big-tech/article/3148128/china-said-suspend-approval-new-online-games-heating-beijings

三个月增长近10倍 "元宇宙" 商标申请总量超8534条. (2021). 证券日报. https://finance.sina.com.cn/tech/2021-12-20/doc-ikyakumx5190942.shtml

丹.声.道. (2014a). 大众点评张涛: 马化腾想明白了, 很脏很累很苦的活腾讯其实不愿意做. 虎嗅网. https://www.huxiu.com/article/38149.html

他距离张小龙就差一个和菜头了. (2017). 界面新闻. https://m.jiemian.com/article/1159854_yidian.html

何苦做游戏:腾讯天美老大姚晓光成长故事-新闻频道-手机搜狐. (2016). Sohu. https://m.sohu.com/n/482986957/?pvid=000115_3w_a

作者:由曦. (2019). 蚂蚁金服. 中国中信出版社.

卢维.兴. (2005). 腾讯全资收购Foxmail 传收购价500万美元. 新浪. http://tech.sina.com.cn/i/2005-03-18/0940554592.shtml

双十一! 4982亿! 还有一些你不知道的数据!. (2020). 微博大数据研究院. https://hd.weibo.com/article/view/2594

吴., & 风. (2016). 腾讯传1998-2016:中国互联网公司进化论. 浙江大学出版社; 第1版.

周鸿祎:真想不通是张小龙这样的人做出了微信! . (2016). Sohu. https://m.sohu.com/n/423292233/?pvid=000115_3w_a

央媒批游戏产业文章重新发布 删除"精神鸦片"等字眼. (2021).凤凰网科技. https://tech.ifeng.com/c/88PSwOgkCrT

宋玮. (2014). 3.0版张小龙: 走出孤独. 财经网. https://m.caijing.com.cn/api/show?contentid=3672661

《广州故事》小黑屋11人 构建出微信的第一个版本. (2019). CCTV. http://tv.cctv.com/v/v1/VIDEFcAfNAVcJ1ZHraG9BHaT190126.html

张小龙-学习笔记. (2019). Allan的笔趣阁. https://www.kancloud.cn/allanyu/reading-room/84147

张小龙: 花儿一样的你, 刺一样活着. (2018, September 29). 新浪. http://tech.sina.com.cn/csj/2018-09-29/doc-ihkmwytp8276956.shtml

微信之父张小龙和他的孤独星球. (2017). 极客公园. https://it.sohu.com/20170929/n515235818.shtml

快手天猫合办双11狂欢夜 4小时直播在线人数峰值超1000万. (2019). 亿邦动力网. https://m.ebrun.com/360397.html

政府工作报告提出"互联网+"概念 马化腾:让人振奋. (2015). 人民网. http://media.people.com.cn/n/2015/0306/c40606-26646256.html

泡泡堂开发商状告腾讯QQ堂侵权索赔50万. (2006). 信报讯. http://tech.sina.com.cn/i/2006-09-11/07351129024.shtml

海南将投数百亿元重点打造琼南琼北两个信息产业基地. (2014). 南海网. http://www.hinews.cn/news/system/2014/10/30/017070490.shtml?wscckey=a96ab2eeb-d9dd264_1633896228

潘., & 王. (2014b). 腾讯方法: 一个市值1500亿美元公司的产品真经 (Chinese Edition) (第1版 ed.). 机械工业出版社.

"独裁"的张小龙和他的微信帝国诞生记. (2013). 博客天下. https://www.tmtpost.com/62285.html

王的男人, 王者荣耀之父, 姚晓光. (2021). 腾讯网. https://new.qq.com/omn/20210505/20210505A057WT00.html

网络游戏的知识产权保护 – "韩国NEXON诉腾讯QQ堂"案引发的思考. (2008). 法律图书馆. http://www.law-lib.com/lw/lw_view.asp?no=8845

腾讯副总裁姚晓光:《王者荣耀》吸引两亿用户的秘诀. (2019). 天美工作室群. https://www.youtube.com/watch?v=eYFS7eCllEE

腾讯第一位社招员工任宇昕的故事: 从码农到腾讯COO. (2017). Game Look. http://www.gamelook.com.cn/2017/10/307161

腾讯首席技术官熊明华:怎样打造创业团队-HD高清. (2013, January 23). 在路上. https://www.youtube.com/watch?v=eDNSTCfbqKQ&hl=id&client=mv-google&gl=ID&fulldescription=1&app=desktop&persist_app=1

西湖论剑物是人非 第三代大侠开场实录. (2002). 新浪科技. http://tech.sina.com.cn/i/c/2002-11-03/1539147650.shtml

身价88亿元 31岁的陈天桥成IT全国首富. (2004). 中国广播网. http://news.sina.com.cn/s/2004-10-21/07453988396s.shtml

陈. (2018). 这就是马云 (马云亲笔作序,马云助理精心撰写,畅销40万册,横扫全球20 多个国家和地区,史玉柱等大咖大力推荐,阿里巴巴太极禅苑火爆销售中) *(Chinese Edition)* (第1版 ed.). 浙江人民出版社.

马化腾　打开未来之门. (2018). 中国民航报社. http://www.caacnews.com.cn/special/4260/4992/ggkf40nlm5/201812/t20181210_1262582.html

马化腾公开信: 移动互联网时代更需要创新. (2014). 新浪科技. http://tech.sina.com.cn/i/2014-10-29/22169744202.shtml

马化腾给他发2亿奖金, 他创造了腾讯25%的营收. (2014). 新浪科技. https://cj.sina.com.cn/article/detail/2041009203/411924?cre=financepagepc&mod=f&loc=1&r=9&doct=0&rfunc=100

马化腾: 腾讯未来会通过康盛投资站长. (2011). 腾讯科技. https://tech.qq.com/a/20110521/000122.htm

INDEX